Hidden Histories
Coventry Jewish Watchmakers

Mark Johnson

Mark Johnson has asserted his right under the Copyright, Designs and Patents Act 1988 to be identified as the author of this work.

Copyright © Mark Johnson 2022-23

Acknowledgements for the source material drawn from is given within the book.

First Edition March 2023

File Vn 718, Updated 13:59 on 14-Mar-23

The book contains the research conducted for https://www.stamproductions.co.uk/ 'Migrant Families | Hidden Histories' including the Coventry Jewish Watchmakers

First published in Great Britain in 2023 by Lean Business Vision

A catalogue record of this book is held at the British Library

ISBN-13: 978-1-7393102-0-2

This book is sold subject to the condition that it shall not, by way of trade or otherwise, be lent, resold, hired out otherwise circulated without the publisher's prior consent in any form of binding or cover other than that in which it is published and without a similar condition, including this condition, being imposed on the subsequent purchaser.

Cover Photos:

Alfred Emanuel Fridlander Gold Watch.

Spon Street, the centre of the Coventry watchmaking trade, looking towards Moss Fridlander's shop in Fleet Street.
From the 1891 directory 'Where to buy in Coventry'

Coventry Watchmaker: Philip Cohen (1823 – 1898)
From 1891 trades directory 'Where to buy at Coventry'
Image Public Domain, Re-colourised by the author

Philip Cohen Silver Watch movement.

Photos by the author

Endorsements

After a lifetime in horology, it was a real pleasure to read such a fascinating book on a topic that, until now, I knew very little about.

Hidden Histories is not only very well researched, relaying stories of what life was like as a Jewish watchmaker in the Coventry area, but is also accompanied by a host of captivating images, which bring the whole book to life.

I have thoroughly enjoyed reading it, I am delighted the story has been told and I hope you enjoy this book as much as I have.

Paul Roberson FBHI, Chairman, British Watch & Clockmakers Guild

As someone involved in providing a welcome to people arriving in Coventry seeking shelter from conflict or various forms of persecution, I cannot stress enough how some of the themes of the book resonate. Some of the challenges remain the same for people having to re-start lives often from scratch, but also many of the positives new communities bring, innovation, entrepreneurship etc also resonate.

I thoroughly enjoyed the book, which really brings Coventry's Jewish community to life through great illustrations, human stories and references to the Coventry we live in today

Peter Barnett CBE, Head of Libraries & Migration, Coventry City Council

Mark Johnson's forensic and beautifully illustrated study of Coventry's Jewish watchmakers and their families is an important contribution to understanding the pre-industrial city's most important trade. It is also a powerful testimony to the benefits of migration, which has always shaped Coventry's story.'

Peter Walters, Chair, Coventry Society

I very much enjoyed reading 'Hidden Histories' and realise how important the research is in filling in the gaps in Coventry's history, especially with regard to highlighting the key role played by immigrant workers in the watchmaking industry. It is fantastic that this is now being acknowledged.

The meticulous research of the immigrant families featured in the book makes for very interesting reading. And the recording of the growth of the Jewish Community leading to the opening of the Coventry Synagogue is a much-neglected part of the city's history

Jane Railton, Author, Coventry Watchmakers' Heritage Trail

Mark Johnson is an expert genealogist. He has used his skills to make the story of the Jewish Watchmakers of Coventry a fascinating account of their lives and times. He has unearthed many nuggets that help to enrich the story of their lives and add to the history of Coventry.

Dr. Martin Been, President Coventry Jewish Reform Community

This book achieves what it sets out to do, and that is to review the Coventry watch industry through the eyes of the large Jewish immigrant population of the City. It offers us a peep through the keyhole into an important part of English watchmaking history which has not yet been fully recorded.

This book will find a place in the library of anyone interested in English watchmaking for the light it shines on the watchmaking district of Coventry.

Chris Papwroth MBHI, TimePiece Magazine

This is a truly excellent and fascinating book about Coventry history, so well-written and researched. A very readable and well-illustrated book.

Jill Prime, Trustee and Heritage Researcher, Coventry Watch Museum

Jewish Watchmakers of Coventry

Contents

Dedication	iii
Foreword	v
Introduction	vii
Watchmakers' Stories	**1**
1. Immigrant Families – Prussia, Bavaria and a Bride from Jamaica	1
2. Isaac & Judy Cohen – Married 78 Years and lived to 107 & 101	3
3. Fair Rosamund	5
4. Moss Fridlander – Catching the Oswestry Murderer	7
5. Where to buy at Coventry – Philip Cohen, Watch Manufacturer	10
6. Cohen, Solomon & Co Watch Manufacturers – Antisemitism	12
7. Alfred Emanuel Fridlander – Entrepreneur & Community Leader	14
8. Commercial Travellers – Levin Joel, Jane Joel & Lewis Anidjah	17
9. A Jewish Wedding in Coventry - Lizzie Baum to Jacob Landau	19
10. Coventry Synagogue	21
11. Coventry Hebrew Congregation – An Astonishing Discovery	23
Watchmakers Family Trees	**25**
12. Coventry Jewish Watchmakers Family Tree Overview	26
13. Cohen, Solomon, Samuel, Fridlander, Klean, Baum & Silveston Family Tree	27
14. Radges, Harris & Joel Families	28
Where they Lived & Worked	**29**
15. Map of Coventry Jewish Watchmaker Sites	29
16. Cambridge Villa, Holyhead Road - Alfred & Flora Fridlander 1861-1901	30
17. Mandeville, Hertford Place - Alfred & Flora Fridlander 1911-29	32
18. Rothesay Terrace, Barras Lane – Marks Baum 1891	33
19. Bayley Lane – Francis & Paulina Silveston 1851	33
20. Butcher Row – Isaac & Judy Cohen 1722	34
21. Fleet Street - Moss and Mary Fridlander 1841-58	35
22. Oxford Terrace, Hearsall Lane – Philip Cohen 1861-81	36
23. Spon Street - Marks Baum 1871-75 & Francis Silveston 1861-98	37
24. The Butts – Radges Family 1868-98 & Solomons 1861	40
Watchmakers Detail Reference	**41**
Coventry Jewish Cemetery	**49**
Other Key Coventry Watchmakers	**51**
25. Summary Overview	51
26. Bahne Bonniksen	51
27. Rotherham and Sons – Founded by Samuel Vale	51
Sources & Acknowledgements	**55**
Index	**57**

Dedication

To all the immigrant families coming to our City today

who will enrich it and bring life & vitality

to its culture & commerce

and so contribute to building its prosperity

as the Jewish Watchmakers did before them

Foreword

It is well documented that, at one time, Coventry watchmakers were making half of the watches in England and that this skilled craft and workmanship led to the development of bicycles and cars in the city. However, this booklet reveals a hitherto hidden history about the Jewish watchmakers in the City who worked harmoniously within this thriving community in the City in the eighteenth and nineteenth centuries.

Making them visible in relation to their work, their civic contribution and the daily fabric of their lives, this booklet together with the film 'Migration Stories | Hidden Histories' brings alive how these Jewish families from Prussia, Poland and even Jamaica worked and worshipped in Coventry which was their city of sanctuary as it is for so many from other countries today. Migration has been a constant of this City and the UK bringing enrichment, skills, and vibrancy to all our lives.

Professor Emerita Gillian Hundt
University of Warwick

"I wish it need not have happened in my time," said Frodo.

"So do I," said Gandalf, "and so do all who live to see such times. All we have to decide is what to do with the time that is given us."
– J.R.R. Tolkien

Introduction

Rev. Nathan Marcus Adler, Chief Rabbi. Photo: Public Domain

The station clock read precisely 12:25 on the 4th September 1870 as the steam train pulled into Coventry station on time. The Very Reverend Nathan Marcus Adler, esteemed Chief Rabbi of the British Empire, stepped from the carriage to be greeted by watchmakers, Alfred Emanuel Fridlander and Philip Cohen, Secretary & President of the newly built Coventry Synagogue.

The party proceeded to the home, and watch manufacturing factory of Philip Cohen at 1 Oxford Terrace, Hearsall Lane in Coventry's expanding watch trade area of Chapelfields, where a sumptuous déjeuner was prepared.

At two o'clock, having dined with some of the Country's finest watchmakers, the Chief Rabbi and his entourage proceeded to the nearby synagogue in Barras Lane, where the assembled throng of the Jewish community, and the good, and great, of the city were assembled.

The Chief Rabbi first examined the children of the Sabbath Class, and expressed his surprise at the proficiency of children who were taught by watchmakers Alfred Fridlander and Selim Samuel. He then conducted the dedication service for the synagogue. He commended the congregation and offered up a consecration prayer, which was followed by a prayer for Queen Victoria, the Royal Family and the government.

The Jewish watchmakers were all immigrant families. They did not pray this final prayer lightly, for they had known for centuries, how fragile was the existence of their community – be it suspicion, insults, accusations, lack of emancipation, restrictions on which trades could be worked, or outright hostility and destruction - as would follow just a few years later in Russia.

Many of the families were from Prussia and Bavaria in today's Germany. In his book 'Bloody Foreigners' which traces the story of immigration to Britain, and the challenges faced by all those communities, Robert Winder highlights that many Jewish immigrants 'had suffered severe antisemitism in Germany.'

Having been excluded from all the mediaeval trade guilds across Europe, it would have been with joy that the Jewish watchmakers found no such limitations when the British Watch and Clock Makers' Guild was established in 1907 by one of Coventry's non-Jewish immigrant watchmakers, Bahne Bonniksen.

The Jewish watchmakers, like so many immigrant communities to this present day, depended on tolerance and understanding overcoming the often-underlying fear of the 'other'. That they established a thriving community which contributed so highly to the life of the city is testament to their creative tenacity and commitment to the wellbeing of the wider community that they now served.

Alfred Fridlander Gold watch achieving Class 'A' Certificate from Kew Observatory. Photo courtesy of Tony Barber

Coventry has had a rich industrial and commercial heritage over the centuries. It has re-invented itself in the wool trade, in weaving, in ribbon making, in watchmaking, in sewing machines, bicycles and car design and manufacture. In each of these fields Coventry has benefitted from the skills of immigrant families. They have pioneered and led the world with quality products admired and sought after around the globe.

Such was the case for the watchmaking trade. Starting from small beginnings in 1747, Coventry grew to be the world centre of watchmaking a hundred years later. Among the pioneering entrepreneur watch manufacturers were several significant Jewish watchmakers. They played key roles in the development of the City, its community and its commerce. It was in the heyday of the Jewish watchmakers that Coventry's Jewish Community grew strongest and its own synagogue.

Fridlander watch movement showing the Class 'A' Certification from Kew Observatory. Photo courtesy of Tony Barber

The book traces the lives, adventures and achievements of these pioneers – all from immigrant families – through the joys and challenges that they faced in the rise, and the fall of Coventry's watchmaking trade.

The **Watchmakers Stories** in chapters 1 to 11 bring to life some of the key characters, telling some of the intriguing stories around their lives.

The **Watchmakers Family Trees** in chapters 12 to 14 describe the intertwined family connections between the Jewish Watchmakers in their Family Tree with marriages between no less than seven of their families.

In **Where they Lived and Worked** – chapters 15 to 24 - we map out their homes and places of watch manufacturing around the Spon Street and Chapelfields areas. These expanded rapidly with the growth of watchmaking in Coventry. This can be used to track your own guided trail recounting their stories in the very places where they occurred.

In the **Watchmakers Detail Reference** section on p.41, we find details of the dates and places of key events in their lives.

Many of the Jewish watchmakers moved on to London and elsewhere as the watch industry changed. For those that stayed, the **Coventry Jewish Cemetery** section on p.49. begins a record of their burials in the London Road Cemetery.

To provide some of the broader context of the development of watchmaking in Coventry Chapters 25 - 27 provide some pointers to the lives of other big players in the development of the industry.

In compiling this book, I am grateful to the Coventry Watch Museum, and to Harry Levine's 1970 book 'The Jews of Coventry' as well as countless other sources in building the rich tapestry that is presented here.

The research for the book was inspired by Claudette Bryanston and her work with Stamp Productions (http://stampproductions.co.uk) in the production of their film 'Migration Stories | Hidden Histories' which captures fascinating and compelling stories of migration to Coventry, from history to the present day and is published in video form.

You will discover as you read, how Coventry's Jewish watchmakers made a major contribution to the life of the City and its prosperity, as well as to the Jewish community itself.

Watchmakers' Stories

1. Immigrant Families – Prussia, Bavaria and a Bride from Jamaica

All of the Coventry Jewish watch manufacturer families were 1st or 2nd generation immigrants. They would have gone through the upheaval and heartache of leaving family roots and establishing home in a new land, and with a new language.

The rapid expansion, … and then decline, of the Coventry watch trade led to further migration both to and from Coventry.

> The Solomon family were from Germany
> The Cohens from Prussia
> The Fridlanders from Bavaria
> The Kleans from Germany
> The Radges from Prussia
> The Silvestons from Poland
> The Baums from Germany
> And Philip, Evelina & Flora Solomon from Jamaica

The Jamaican connection is a delightful one, expressing the home-longing of immigrant families:

> Alfred Fridlander's wife Flora Sarah Solomon was born in Mandeville, Jamaica. Her mother Evelina Solomon née Brandon was born there, the daughter of Abraham & Judith Pinto Brandon. Evelina and Philip Solomon were married there - a cross-community Sephardi-Ashkenazi wedding!
>
> Alfred & Flora were married by the Chief Rabbi Dr. Nathan Adler, at the luxurious Willis Rooms in London. Alfred & Flora's final home was in Hertford Place and was called 'Mandeville' to remember her roots - carrying the ties to her homeland throughout her life. See Chapter 7 for more of their story.

Like many immigrant communities, the Jewish Watchmakers faced misunderstanding and racism from those around them, despite all their endeavours to contribute to the life of the City.

> Moss Fridlander served the City as a Councillor, but his application was opposed by some on the basis of him 'being an alien', with him having to provide evidence of his Naturalisation Papers as his right to serve – 'Guilty, unless proven innocent'.
>
> The same fear of 'the other' fed the widespread antisemitism faced by the Jewish community everywhere – In 1893 the Coventry Herald and Free Press printed a rebuttal by the Chief Rabbi Dr. Herman Adler which you can read in Chapter 3.
>
> A blatant case of antisemitic racism against Philip Cohen & Philip Solomon's watch manufacturing business was published in 1862 by both the Coventry Herald and the Coventry Times. It is heartening to see the defence published a week later in a letter signed by 63 of their, mostly non-Jewish, employees. See the full story in Chapter 6.

All this despite the many ways that the Jewish Watchmakers served the City, these went well beyond contributing to its prosperity and employment:

> Alfred Emanuel Fridlander served as Honorary Secretary to Coventry's Volunteer Fire Brigade; he served as a City Councillor; contributed to many charitable works; and served as a Justice of the Peace.
>
> Integration with the life of the City was beautifully expressed when Moss Fridlander took the initiative and went out of his way to collaborate with the City's Chief of Police that led to the capture of The Oswestry Murderer. See Chapter 4 for the full story.

Commerce and industry were very dynamic in Victorian times. The watch trade grew rapidly during the mid-1800's bringing much immigration – both business developers, and workers to support them. By the late 1800's, however, cheap imports from America and Switzerland saw the rapid decline of the watch trade – with skills being switched to pioneering and manufacturing sewing machines, then bicycles and then cars.

Some stayed and transitioned to the new industries. Many moved on to London or elsewhere. The following table and chart show the migration of the key Jewish Watch Manufacturers to, through, and from Coventry …

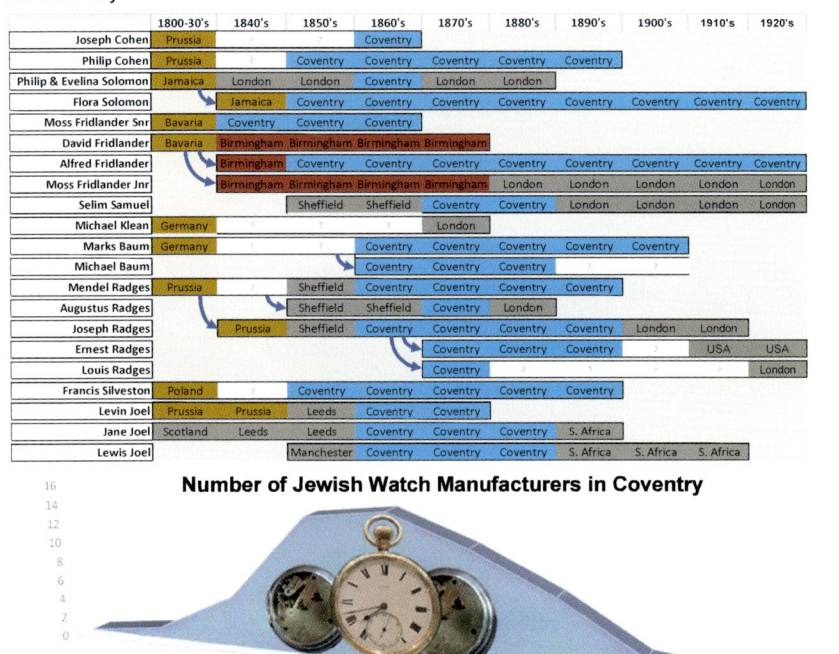

	1800-30's	1840's	1850's	1860's	1870's	1880's	1890's	1900's	1910's	1920's
Joseph Cohen	Prussia	?	?	Coventry						
Philip Cohen	Prussia	?	Coventry	Coventry	Coventry	Coventry	Coventry			
Philip & Evelina Solomon	Jamaica	London	London	Coventry	London	London				
Flora Solomon		Jamaica	Coventry	Coventry	Coventry	Coventry	Coventry	Coventry	Coventry	Coventry
Moss Fridlander Snr	Bavaria	Coventry	Coventry	Coventry						
David Fridlander	Bavaria	Birmingham	Birmingham	Birmingham	Birmingham					
Alfred Fridlander			Birmingham	Coventry	Coventry	Coventry	Coventry	Coventry	Coventry	Coventry
Moss Fridlander Jnr		Birmingham	Birmingham	Birmingham	Birmingham	London	London	London	London	London
Selim Samuel			Sheffield	Sheffield	Coventry	Coventry	London	London	London	London
Michael Klean	Germany	?	?	London						
Marks Baum	Germany	?	?	Coventry	Coventry	Coventry	Coventry	Coventry		
Michael Baum				Coventry	Coventry	Coventry	?	?		
Mendel Radges	Prussia	?	Sheffield	Coventry	Coventry	Coventry	Coventry			
Augustus Radges			Sheffield	Sheffield	Coventry	London				
Joseph Radges		Prussia	Sheffield	Coventry	Coventry	Coventry	Coventry	London	London	
Ernest Radges				Coventry	Coventry	Coventry	?	USA	USA	
Louis Radges				Coventry	?	?	?	?	London	
Francis Silveston	Poland	?	Coventry	Coventry	Coventry	Coventry	Coventry			
Levin Joel	Prussia	Prussia	Leeds	Coventry	Coventry					
Jane Joel	Scotland	Leeds	Leeds	Coventry	Coventry	Coventry	S. Africa			
Lewis Joel			Manchester	Coventry	Coventry	Coventry	S. Africa	S. Africa	S. Africa	

Number of Jewish Watch Manufacturers in Coventry

As is common throughout time, the watchmaking immigrant families' presence, both Jewish and non-Jewish, was most felt in the poorest areas - in the heart of the Spon Street area and in predominantly the lowest cost housing. Spon Street, Barras Lane and The Butts appear throughout the stories here.

The watchmaking business led to the City's expansion - first to Chapelfields, where Philip & Priscilla Cohen's home and factory are to be found. Just a few years later, Earlsdon started to be built. Here by 1861 over two thirds of the working residents were in the watch trade – but almost without exception, were upgrading from elsewhere in Coventry (44), or from the other watchmaking centres of Prescot (12) & London (3). Even the small number of those whose businesses grew largest still lived in terraced or semi-detached homes – and those had their workshops and factories directly attached. The vast majority lived in the smallest accommodation imaginable – as can be seen at Coventry Watch Museum. Yet the expansion into Earlsdon shows the upward social movement that the immigrant families struggled and strived for as they contributed to the development and prosperity of the City.

2. Isaac & Judy Cohen – Married 78 Years and lived to 107 & 101

Isaac & Judy Cohen were the first of the 'Old Coventry Jewry' and ancestors of Philip Cohen. Isaac was born about 1728 and Judy about 1732. They were married around 1755.

The Jewish Chronicle article on 5th June 1936 records Isaac Cohen as having been a Sergeant Major in the Prussian Army during the War of Frederick the Great – probably in the Silesian Wars between Prussia and Austria 1740-1763.

Isaac and Judy moved to Coventry about 1775. According to Benjamin Poole's 1852 History of Coventry they lived in an obscure little tenement on the North side of a passage leading from Great Butcher Row to Priory Row.

This was the first place of Jewish worship in the City. Gatherings later moved to Derby Lane, Fleet Street and Court 16, Spon Street before the synagogue was built.

Isaac's occupation in Coventry was that of Trunk Maker.

Harry Levin, in his book 'The Jews of Coventry' (1970), writes about them extensively, providing references which he felt assuredly confirmed that they were the parents of Philip Cohen, Coventry watch manufacturer and President of the Coventry Hebrew Congregation from 1850-1853. Whilst this may appear implausible from Philip's birth date of 1827, Harry Levine goes to lengths to convince the reader of its truth.

The Jewish Chronicle article of 5th June 1936 also reinforces this ... although Harry Levine was the Birmingham correspondent for the Jewish Chronicle, so its concurrence of the story is hardly surprising!

A perhaps more likely link is via one Joseph Cohen, a Coventry Watch Manufacturer born 1798 in Prussia, and in 1861 residing at 54 Spon End, Coventry. Joseph died in 1868 in Coventry. No other references to this Joseph Cohen have been found. He certainly was one of the Coventry Jewish Watchmakers and could, perhaps, have been the son of Isaac Cohen and the father of Philip Cohen ... but for now we will let the published version of the Cohen family history stand even though it would seem to need a miracle of reproduction of a truly Abrahamic scale.

Potentially in addition to, Philip Cohen, Isaac and Judy had a daughter known by her Stage Name of "Fair Rosamund" from one of the parts she played – See Chapter 3 for her story.

When Judy died in 1833 aged 101 she and Isaac had been married for 78 years.

> DEATHS.
> On Wednesday, at the advanced age of 101 years, Judy, the wife of Isaac Cohen, of this City.

Judy Cohen's death aged 101 recorded in the Coventry Herald and Observer 5th April 1833

On 1st May 1835 the Coventry Herald and Observer reassured its readers that reports in the London papers of the demise of Isaac Cohen were premature and that he had heartily eaten his breakfast that morning:

> THE PATRIARCH COHEN.—Among the deaths recorded in some of the London papers, we have observed it stated that this lot has fallen to Isaac Cohen, of this City, aged 106. We beg to say that the venerable old man is as well to-day as he has been for some time, and eat a hearty breakfast this morning.

Isaac lived on to the incredible age of 107!

His death was reported in the Coventry Herald and Observer on 18th December 1835 and his eulogy honoured his devout life of faith.

> A PATRIARCH.—On Sunday last, died the Patriarch, ISAAC COHEN, at the advanced age of one hundred and seven years: and for upwards of sixty, an inhabitant of this City. This venerable old man was in the strictest sense of the word a Jew, devotedly attached to his religion, and a scrupulous observer of all the ceremonies of his forefathers. By those who have daily witnessed his habits of living, he was considered a devout man according to his faith, though not always under the most favourable circumstances. His situation in life was humble, but he was respected by all who knew him. About eight or ten years ago, we believe the late Mr. James Grant represented his case to Mr. Goldsmid, the money dealer, of London, who has ever since allowed a sufficiency to preserve him from poverty. He retained his mental faculties to the last moments of his existence, and died repeating the sublime Hebrew prayer, "Hear, O Israel," &c. His remains were conveyed to Birmingham for interment. His wife, with whom he had been united 78 years, died about two years ago, aged one hundred and one.

The following week they corrected their reference to his living allowance which had been from Mr. Rothschild.

> ISAAC COHEN.—In our paragraph last week respecting the death of this venerable old man, we stated erroneously that he had received an allowance for some years past from Mr Goldsmidt: it should have been, from Mr. Rothschild.

Philip Cohen, whether the son, grandson, or other relative of Isaac Cohen, went on to become President of the Coventry Hebrew Congregation and gets his own chapter in the book The Jews of Coventry. Philip Cohen was one of Coventry's leading watchmakers. His home and workshops can still be seen on the corner of Hearsall Lane and the Allesley Old Road.

Philip & Priscilla Cohen's Home & Workshops in Oxford Terrace

Philip Cohen Pocket Watch Courtesy of Leonard Heanes at Coventry Watch Museum

3. Fair Rosamund

The daughter of Isaac & Judy Cohen, the patriarchs of Old Coventry Jewry, was known by her stage name "Fair Rosamund". She performed in Loew's Dancing Show at Coventry's Great Fair. The Great Fair took place in June each year.

In 1832 she played the part of Britannia in a procession in celebration of the Reform Act. The Act had introduced major changes to the electoral system giving representation to cities, and the vote to small landowners, tenant farmers, shopkeepers, and some householders lodgers. People from all walks of life came out to celebrate.

Intriguingly, sixty years later, on 10th Feb 1893, the Coventry Herald and Free Press records Coventry's great star of the theatre, Ellen Terry, as playing "Fair Rosamund" in Mr. Henry Irving's production of Lord Tennyson's play "Becket" - The article is immediately followed by a defence against antisemitism by the new Chief Rabbi Dr. Herman Adler, son of Reverend Nathan Marcus Adler, who had dedicated Coventry Synagogue:

In certain political circles no language is held to be too strong for the denunciation of the alien Jew immigrant. But if what Dr. Adler, the Chief Rabbi, has been telling an interviewer is correct, the alien Jew immigrant is not as black as he is painted. To be sure, Dr. Adler is not an altogether impartial witness. He is the co-religionist of the poor people so violently attacked, and it is his duty to care for their spiritual interests. But he stands as high in public estimation as does the chief of any other denomination in this country. Dr. Adler explains that the Jewish immigration comes almost exclusively from Russia, and that, despite the outcry raised against it, it is comparatively small, besides being counteracted by a steady emigration to America. The Chief Rabbi contends that they are law-abiding and industrious, and, in the course of time, become good citizens. The children, moreover, show remarkable adaptability to English ways. With no more than two years' training at school they speak and write the language as though to the manner born.

The cheap labour of the East-end Jews has, Dr. Adler contends, been a blessing to the working classes of London. Formerly the British labouring man was seldom able to appear in new clothes. He had to content himself with second-hand ones; but, thanks to the immigrant Jews, tailoring is now so cheap that the second-hand clothing trade is dying out. Cheap bootmaking is also a distinctly Jewish trade, and one of great value to the community. On the sore point of the burdens of the ratepayer, Dr. Adler has also a good word to say for his co-religionists. The Jews make a point of seeing as far as possible after their own poor, and the pauper Jew hardly comes upon the rates at all. Occasionally a deserted wife goes into the workhouse; but in London at the present time there are only twelve such cases. The one subject upon which the Chief Rabbi speaks with bitterness is the Russian persecution of the Jews, which he regards as a fearful wrong, and a disgrace to civilisation. He thinks the time will come when the great Powers will intervene to stop the cruelties habitually and gratuitously practised within the dominions of the Czar.

DR. ADLER, CHIEF RABBI.

4. Moss Fridlander – Catching the Oswestry Murderer

One of the earliest Jewish Watchmakers in Coventry was Moss & Mary Fridlander – already well established in Fleet Street, Coventry by 1841. His shop would have been close to what is now the entrance to the Lower Precinct.

Moss demonstrated highly commendable eagerness to work together with the authorities to bring about justice when his jeweller's shop was approached by a suspicious character in December 1841. His insights, and quick-thinking lead to the capture of the Oswestry Murderer.

News of his actions spread far and wide. Here is the detailed account recorded in the national newspapers of the day...

Moss Fridlander, Evidence to the 1841 Oswestry Murder Trial

London Evening Mail - Friday 31 Dec 1841 – The Chirk, Oswestry Murder of Emma Evans

https://www.britishnewspaperarchive.co.uk/viewer/bl/0001316/18411231/048/0007

Also in the London Evening Standard, the London Morning Post, The London Evening Chronicle, The London Globe, The Shipping & Mercantile Gazette, The London Sun, and the Coventry Standard

> **THE OSWESTRY MURDER.**
>
> **APPREHENSION AND EXAMINATION OF THE MURDERERS.**
>
> (From the *Shrewsbury News*.)
>
> It is scarcely possible to describe the excitement displayed in this town, and indeed in the whole county, by the arrival of the intelligence on Monday, that Williams and Slawson, the two men who are suspected of having committed the brutal murder of Emma Evans, at Bronygarth, near Chirk, had been captured at Coventry; and superintendent Mac-

...

> T. H. Prosser.—I am chief constable of the city and county of Coventry. In consequence of information I received, I apprehended the prisoner John Williams at the house of Mr. Moss Fridlander, a watchmaker and silversmith, residing in Fleet-street, Coventry, on Saturday, December 25. I knew his object, having received information. I saw him enter the shop about half-past 11 o'clock, and enter into conversation with Mr. Fridlander. I was in the back parlour, and saw him come into the shop alone. He was standing with his back towards me when he was conversing with Mr. Fridlander. I did not hear the conversation. I was looking through a window which looks into the shop. After a short time I opened a door which led to the shop, and said, "Hallo, Williams!" He turned round and looked at me. I said, "I think you are the man I want, John Williams; that's your name, is it not?" He said "Well, what of that?" I pulled a pair of handcuffs out of my pocket, and said to him, "I must just put these on. You come from Wrexham, don't you?" He replied, "Well, what if I do?" I took him into custody. I then

About 7 o'clock on Thursday evening, the 16th December 1841, the body of Emma Evans, an old maiden shopkeeper, was found dead upon the middle of her kitchen floor, in a stream of blood, with her pocket turned inside out. She lived near the village of Chirk, 5 miles from Oswestry in Shropshire. When it came to trial ...

Mr T. H. Prosser reported to the court: I am the chief constable of the city and county of Coventry. In consequence of information I received, I apprehended the prisoner John Williams at the house of Mr. **Moss Fridlander, a watchmaker and silversmith, residing in Fleet Street, Coventry**, on Saturday, December 25.

I knew his object, having received information. I saw him enter the shop about half past 11 o'clock, and enter into conversation with **Mr. Fridlander**. I was in the back parlour, and saw him come into the shop alone... I was looking through window which looks into the shop. After short time I opened door which led to the shop, and said,
"Hallo, Williams!" He turned round and looked at me.
I said, "I think you are the man I want, John Williams; that's your name, is it not?"
He said "Well, what of that!"

I pulled a pair of handcuffs out of my pocket, and said to him, "I most put these on you. You come from Wrexham, don't you?" He replied, "Well, what if I do?"

And I took him into custody.

I then said, "Where's Slawson, is he outside?" He said "No." I then said, "When did you see him last?" He said, "Yesterday morning." I said, "Where?" "In the street," he replied.

"Do you know where he is?" "I do not," he said "but I believe he has got work."
I said, "Do you know where?" He said, "I think the name is Lanshaw."

I then searched him in the shop. He allowed himself to be searched willingly.

In his coat pocket I found a German silver caddy spoon, or sugar scoop, with no mark (but a stamp in imitation of silver), knife in case (used for cutting pigs), and knife with horn handle, which had been recently sharpened on stone. In the inside pocket of his waistcoat I found a paper. The paper was here read, and contained the following advice:

> "You most make the best of your way through South Wales to the Bristol Channel, and cross over to Cornwall, and inquire at Tregonna for some of the Beards, if you are lucky enough to get there."

Mr. Prosser then continued his examination as follows: — I found two old dirty pocket handkerchiefs, comb, a purse with worsted and cotton in it, copper nail, a tobacco-stopper, and button. From **Mr. Fridlander**, in the presence of Williams, I received one silver teaspoon (broken in three parts), with the initials "**E.E.**" in Roman characters, ten teaspoons of the same description, and pair of sugar-tongs, marked the same as the spoons.

I said to **Fridlander**, "I'll take the spoons," and I received them in the presence of the prisoner. ...

Mr. Fridlander, of Fleet-street, Coventry, silversmith and watchmaker, was next examined, and stated follows: — On Thursday the 23rd inst., the prisoner came into my shop in the afternoon, and asked if I bought old silver? I replied in the affirmative, and then he pulled out a teaspoon, broken in three pieces. I said to him, "I never buy plate, unless I know the parties." He said, "I am no stranger, I live close by, in Spon-street". I said, "What's your name?" He answered, "Edward Jones." I asked if he was a housekeeper, and he said, "Yes; but on account of the badness of trade, I am obliged to sell this."

I weighed it, and it came to 2s., at the rate of 4s. 9d. per ounce; and I paid him for it. I put the spoon in a drawer by itself.

Friday, in the afternoon, the prisoner called again at my shop and pulled ten spoons out of one pocket, and pair of silver sugar-tongs. He laid them on the counter, and said, "I am compelled to sell them: it is Christmas-time, and I have no money."

I had suspicion of him, and said, "What's your name?" and he said, "Edward Jones, and I live in Spon-Street." I said, "Then the spoons are not yours?" He said, "Yes, they are."

Your name is not on them," I replied. He said, "No, not mine, but Edward Edward's."

I then said, "I suppose you got them from some relation?" and he replied, "Yes, they were gift." I then weighed them, and they came to £1. 5s., but having my suspicions, I pretended that I had no change, and told him to call again in half hour.

I did this that I might make inquiries. I then went to a person named Smith, an engraver, and afterward to Mr. Prosser, the chief-constable. Mr. Prosser was away from home, and it was uncertain when would return, so I went home.

Very soon after, the prisoner to my house, and I told him I had not been able to obtain change, but if he would come in the morning. I could pay him. He asked if I could let him have a trifle then, and I gave him 3s.

Next morning, he came at 9 o'clock, and I told him had come too soon, and that he must call again. I then went to Mr. Prosser, and he came to my house, and I put him in the back parlour.

The prisoner came into the shop a little before 12 o'clock, and said,
"Have I come right now?" and I replied, "Yes: quite right."

I then locked the front door, as I had been directed by Mr. Prosser,
and he took the prisoner into custody.
I gave the spoons and sugar-tongs to Mr. Prosser,
and they the same he has now produced.

Epilogue:

In March 1842 John Williams was convicted of the wilful murder of Emma Evans and sentenced to death.

Joseph Slawson was acquitted of the murder, but convicted for robbery, and sentenced to 7 years transportation to Tasmania.

(See Staffordshire Advertiser 26 March 1842).

> Records show that Prisoner number 7254 Joseph Slawson aged 22 was convicted on 18 Mar 1842 of Larceny. He was transferred to the Prison Hulk Ship Justitia, moored at Woolwich on 15 Apr 1842. He was transported later in 1842 on the ship Waterloo from Sheerness via the Cape of Good Hope to Port Arthur in Tasmania for 7 years.
>
> Prisoners at Port Arthur were subjected to an extraordinarily harsh regime. There is no record of his return – very few of them did.

Moss Fridlander's suspicions had been correct. His actions in collaborating with the Coventry Chief of Police had led to the capture of the murderer and of his accomplice.

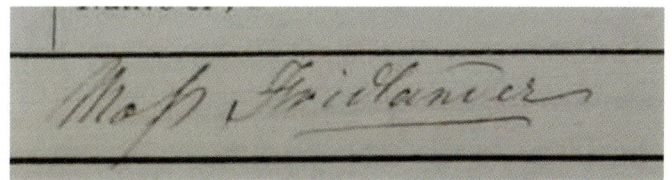

Fridlander, Moss Signature on his July 1841 Port of London Certificate of Arrival

5. Where to buy at Coventry – Philip Cohen, Watch Manufacturer

Most trade directories provide an index to traders, giving just the name of the proprietor and their trade and location. A notable exception is the delightfully titled 1891 directory:

> "Where to buy – An Illustrated Local Traders Review by the Editor of 'The Agents Guide' – The Premier Shops, Manufacturers & Retailers, with Illustrated Descriptions of The Midlands and the West of England – COVENTRY".

Yes, all that is just its' title! Among the 'Premier Manufacturers' is Philip Cohen, watch manufacturer. He is given a full page spread and portrait, complete with a tour of his factory and a detailed description of the working environment and how he provided well for those who worked there. Here is the full account:

WHERE TO BUY AT COVENTRY

Mr. Philip Cohen, Wholesale Watch Manufacturer, Oxford Terrace, Chapel Fields.

To those at all intimately acquainted with the trade of Coventry, the important place held by watchmaking is well known. Owing to the nature of the trade, very little exterior show is made even by the largest makers; so little indeed, that the uninitiated stranger might visit Coventry a number of times without knowing that he was in the chief centre of this branch of industry in England, particularly as the trade is mainly confined to one part of the City

The actual number of firms engaged in this line is, however, surprising, considering that the demand for high-class watches must, from their expensive character, always be more or less limited; and among them all, the claims of Mr. Philip Cohen to be considered a representative manufacturer will be readily allowed.

Both the extensive nature of his business and the superior quality of all articles bearing his name bear this out, whilst he himself is, undoubtedly, one of the best-known men in his line.

The portrait we publish above will be readily recognised by our readers as that of Mr. P. Cohen, and the following account of his factory, and the business conducted therein, will prove, we hope, interesting.

Progress marks the spirit of the age. In every branch of manufacture this is observable, and in none, perhaps, more than in the watch trade.

The principle adopted in one watch is doubtless analogous to those which marks another, but so many modifications and improvements have been introduced, that a watchmaker of a hundred years ago would probably be at a loss to understand, at first sight, some of the parts in a first-class modern watch.

Mr. Cohen is eminently a modern watchmaker, and his productions represent the newest developments in the art of watchmaking; and not only so, but he has himself, in many instances, been the pioneer and introducer of improvements in the now almost perfect timekeeper.

The business was established in 1846, and is now, as it has been from the start, among the leading ones in its line in the country.

To successfully cope with a trade so large as that carried on in this establishment, and give working room for the sixty of so inside-hands – that number being employed as well as many out-workers – requires premises of considerable extent. These have been provided by the erection of large buildings, about forty yards long, containing all the workshops, show-rooms, packing-rooms, and offices needed, the whole being planned with a view to comfort, utility, and the convenience of the proprietor and his subordinates, and to the speedy carrying out of all orders.

The factory is extremely well fitted throughout, with mechanical appliances, and is warmed by excellent steam heating apparatus, besides being well lighted and perfectly ventilated. Recently, the proprietor has added very commodious lavatories, and out offices for the convenience of his workpeople, whom he endeavours to make as comfortable as possible during the time they spend in the factory. These works are attached to the residence of the manager, who can thus exercise a constant supervision over the whole establishment.

All kinds of the best classes of gold and silver keyless lever watches are manufactured in these works, while many of the more expensive kinds, such as chronographs and chronometers, are made to order.

The chief speciality is the making of the well-known "Kew watches". The most extraordinary care is taken to ensure the accurate time-keeping of these watches, each one being subjected in the factory to the rigid test for a lengthened period, and the slightest variation noted. After they are sent to Kew, and, if found absolutely correct, they receive a certificate from the Kew observatory before being delivered to the retailer. With all these precautions, a "Kew watch"[1] must be as perfect as possible for human skill and ingenuity to make it, and they are consequently in great request.

The work done in the factory is divided into separate departments for convenience and better conduct of the various processes needed in the manufacture. There are special rooms for case-springing, polishing, testing, and the manufacture of interior parts, which is only entrusted to the most experienced hands.

One of these rooms is twenty-five yards in length, and well lighted all round. A suitable room is also set apart for engine turning and engraving the backs of the cases. This is done by a machine which was exhibited at the Coventry and Midland Art and Industrial Exhibition [1867], when the judges awarded Mr. Cohen prize medal for his watches.

The foreman of the engine-turning department gives a very interesting account of his experience when exhibiting the working of the machine. He says among other things, that the curiosity and interest excited was so great that people not only flocked all round him, but also actually got on his back to obtain a good view. This was probably awkward, but, from appearances, we should think that his back was well able to bear the burden.

Some forbearance must also have been required and extended, as this gentleman's countenance bore the impress of good nature and health – "features," by the way, that were very common among the operatives in Mr. Cohen's factory.

Although we observed at the beginning of this article. Watchmaking varies little in the actual processes of manufacture, there is, as we all know, a vast difference in the value of articles turned out by different firms. Being so delicate a construction, the slightest irregularity or discrepancy will very soon make itself known in the time-keeping powers of a watch.

On this the "going" qualities of the article, a firm's reputation depends, and accordingly every manufacturer of repute is scrupulously exacting as to the quality of the work that bears his name.

Mr Cohen's watches bear a high reputation in the trade, and amply justify the position he has gained as a wholesale manufacturer

---

[1] Philip Cohen not only regularly achieved Kew Certificates for his watches, but in 1893 appeared in the Royal Society, Kew Observatory Annual Report of the best performing watches in the Country.

6. Cohen, Solomon & Co Watch Manufacturers – Antisemitism

As with communities all around the world, hostilities were encountered by the Coventry Jewish Watchmakers.

One such incident was the unwarranted 'Comparison' of the generosity of a gentile Coventry business with the apparent lack of such, by a business of another faith.

This came about – and was suitably rebuffed – as follows.

In 1862 the Jewish Watchmaking partnership of (Philip) Cohen, (Philip) Solomon & Co are slurred in the Coventry Standard, being accused of forcibly deducting charity contributions from their workers' 'miserable wages'.

The following week, they published a gently worded response leaving it for 63 of their employees to write their own letter strongly defending their Jewish employers, commending them for paying fair wages and enabling the voluntary giving to the Poor Fund by their employees.

See the following newspaper article for full details.

Philip Cohen
Photo: 'The Jews of Coventry'

Coventry Standard - Friday 05 December 1862
https://www.britishnewspaperarchive.co.uk/viewer/bl/0000683/18621205/119/0004

A CONTRAST.

— We understand that Mr. Hennell, ribbon manufacturer, has generously undertaken to pay 4s. per week, up to Christmas, to those of his hands who are without employment. He adopts this course in lieu of giving a subscription to the relief fund.

— In an opposite spirit, we have heard of a watch manufacturer, in Chapelfields, — one, certainly, who does not parade Christianity as his religion, who has resolved to stop two-and-a-half per cent. out of the miserable wages of his workpeople, and to appropriate the money thus arbitrarily obtained, a contribution to the relief fund.

Poor and miserable as the recipients of the fund are, we believe they will one and all repudiate participation in any such paltry exaction.

Coventry Times - Wednesday 17 December 1862

AN UNWARRANTABLE ATTACK

Having been made in the columns of the *Standard* upon the firm of Messrs. Cohen, Solomon, and Co., watchmakers, of Chapelfields, Coventry, we consider we are only doing strict justice to the injured individuals by giving the widest publicity we possibly can to the following denial, which has since been addressed to our Friday contemporary:-

Oxford-terrace, December. 1862.

To the Editor of the Coventry Standard.

Sir. - With reference to the paragraph in your paper of Friday last, headed "A Contrast," in which the conduct of a certain ribbon manufacturer compared with that of watch manufacturer, in Chapelfields, who had resolved to stop 2½ per cent, out of the miserable wages of his workpeople, a contribution to the Relief Fund, we beg to observe that, ours is the only firm in Chapelfields who receive weekly subscriptions from their workpeople for the purpose stated, and from the reference you make to our religion, it is obvious we are the watch manufacturers alluded to.

Your observation, in reference to our religion, we accept as a compliment; but that applying to our business, being a most infamous and malicious falsehood, we adopt the only mode at our command to refute, handing you for publication the enclosed document, signed by sixty-three of our workmen, which shows most distinctly that the subscriptions are not stopped as, stated in your paragraph, but that they are the free and voluntary offerings of the men themselves, and cheerfully adopted a medium of assisting their distressed townspeople, who are not so fortunate to obtain employment, which will be added our own subscription. The same document also shows whether the wages we pay are miserable or not; and if your informant requires more convincing proof, we invite him station himself at our factory door, and enquire of the workmen as they pass.

    We are, Sir,
        Your obedient servants.,
           COHEN, SOLOMON, Co.

Coventry, December, 1862.

We, the undersigned workmen, in the employ of Messrs. Cohen, Solomon, and Co., Chapelfields, having had our attention drawn to paragraph in the Coventry Standard, of the 5th inst, 'A Contrast.' which we believe (from the fact of our having contributed to the Relief Fund in the manner there mentioned) is directly applied to our employers, feel it our duty, in Justice to them, to state that, so far from the amount we contribute being ' arbitrarily stopped from our wages,' we were kindly told by them that if felt disposed to contribute to the Relief Fund, they would be the medium of collecting our contributions; the subscription was, therefore, quite voluntary on our part, and most emphatically deny that it was, or is in any way, compulsory.

We also deny that the wages receive are described in the paragraph, 'miserable,' we feel convinced that we are receiving fully as high prices as those paid by every other respectable house in the trade, and with which we are quite satisfied:

    Wm Huhne Charles Mackanally William Giggins John Ryley Thomas Bayes John Charles Phipps Joseph Frearson James Humphreys J. Davoile David Barnacle Thos. G. Furneaux Samuel Randle John Overton John Price Martha Bridge Samuel Bead Joseph Gore William Payne John Parner Joseph Lynes David Steadman Henry Pavne George Atkins George Henry Hopkins Edward Clarke Benjamin Audley Samuel Ward Joseph A. Atkins Wm. H. Sandes, jun. Cornelius Arnett William Felton Henry Arnold Samuel Horne Henry Collingbourne Eusebius Tuckey Thomas Flowers Thomas Webb Jas. Evans William Glover D. Hannah John Bowen Thomas Oates Robert Tipping Harrison George Simms J. Aviss William North James Lee
    William Lines William Bales Thomas Garner Dalton William Bennett Charles Iliffe John Parsons John Edwin Walton John Gilbert Thos. Pearman James Byron Edward Beaumont George Underhill William Gerrard

7. Alfred Emanuel Fridlander – Entrepreneur & Community Leader

Alfred Emanuel Fridlander was born on 12 June 1840 at 134 Digbeth, Birmingham to David Fridlander & Esther née Emanuel. His father, David, was a Pawnbroker and, later, a silversmith from Bavaria.

On 3rd June 1863 Alfred Fridlander married Flora Sarah Solomon in the luxurious Willis Rooms, also known as the Almack's Assembly Rooms, King Street, St. James, Marylebone. The wedding was conducted by the Chief Rabbi Dr. Nathan Adler.

Marriages.

On the 3rd inst., in London, (by the Chief Rabbi, Dr. Adler), Alfred, eldest son of David Friedlander, Esq., of Kensington-place, Bristol-road, Birmingham, to Flora, only daughter of Philip Solomon, Esq., both of Coventry

Coventry Times 10th June 1863

Flora was born in the town of Mandeville, in Manchester Province, Jamaica. Flora and Alfred lived much of their married life at Cambridge Villa, Holyhead Road, Coventry. They later moved to 35 Hertford Place, off Queen's Road, now adjacent to the Ring Road. Their home was called "Mandeville" after Flora's original home in Jamaica.

Alfred built up a large watch manufacturing business in Coventry, employing upwards of 70 workers. He served on the City Council, on many community groups and as a Justice of the Peace. He was generous in supporting many endeavours, for example donating a silver watch, valued at £4 10 shillings, as the prize in the City's swimming competition in 1864.

Fridlander Watch Face 51965

SWIMMING MATCHES AT THE BATHS.

The second of the annual swimming matches in connection with the Coventry Swimming Club took place at the Baths, Hales-street, on Monday evening. A large number of persons, including many influential members of the City Council, assembled to witness the proceedings. The first contest was for a glass tankard, ornamented with silver filagree work. It was a race between boys, and the following were the competitors:—Benjamin Wilkins, John Hickling, George James, L. Kelly, John Ward, N. Smythe, and Percy G. Roby. The distance was six lengths of the bath, or about 250 yards. The first heat was between Hickling, Wilkins, and James, and was won by about a yard by Wilkins, Hickling being a good second. The other three then jumped in, and Ward won easily. The deciding heat between Ward and Wilkins took place at a subsequent period of the evening, and was won by Ward by four yards. The next match was for a substantial silver watch, value £4. 10s., presented by Councillor Fridlander. There were three competitors—Thomas Robinson (of Birmingham),

Alfred Fridlander Silver Watch Prize in Swimming Competition - Coventry Standard 27 Aug 1864

Alfred Emanuel Fridlander - Photo from the book 'The Jews of Coventry' Restored and colourised by the author

14

Alfred Fridlander produced watches of exceptional quality and renown. The standard for watches was established by the Royal Society Observatory at Kew. They conducted a trial Watch Rating process in 1883 – although from their report of proceedings it would appear that this consisted of little more than that they had 'fitted up at the observatory a first-class burglar- and fire-proof safe for the safe custody of watches'!

The standards were developed to measure the precision of time keeping and the resilience of that accuracy with movement of the watch and with temperature change.

Results of the tests were reflected in individual certification, with the Results of Watch Trials being published in the Society's annual report for those achieving the highest marks. This league table first appeared in 1886. A. E. Fridlander watches started appearing in this prize list from the following year.

A Fridlander watch was 25th in the table in 1887; rising to 13th in 1888; then achieving the first of several Top 1st Place awards in 1889 for the most accurate watch in the Country.

Fridlander Watch with Gold Case in the Herbert Museum, Coventry

> In Appendix II will be found a table giving the results of trial of the 51 watches which gained the highest number of marks during the year. The highest place was taken by Mr. A. E. Fridlander, of Coventry, with the keyless going-barrel Karrusel lever watch, No. 25,582, which obtained 90·1 marks out of a maximum of 100.
> This is the first English lever watch to reach the 90 marks limit, and its performance is the best since 1892.

The 'Kew' National Physical Laboratory Report for the year 1900 Rating of Watches and Chronometers

Alfred continued to drive the quality of his watch manufacture amidst stiff and growing competition. He retained Top 10 places every year for the following 12 years, including three 1st place awards. These culminated in his award of 'Best ever English Lever Watch' in 1900 by what had by then become The National Physical Laboratory. This was for his keyless going-barrel Karrusel lever watch serial number 25,582 – the first English lever watch ever to achieve the 90 marks quality standard.

During this period, Kew awarded 577 English watches the honour of publication in their annual report. Of these, well over half (320) were from Coventry – far outstripping London. Head and shoulders above all the other manufacturers in the Country was Alfred Fridlander, with no less than 107 watches in the Kew reports as well as 18 other complex timepieces. None of the London watch manufacturers exceeded 36 such Kew certificates in the same period.

Alfred Fridlander's watch manufacturing achievements were truly world leading, and spearheaded the growth of the watchmaking industry throughout the Country.

Fridlander watch movement still spinning after 100 years of use

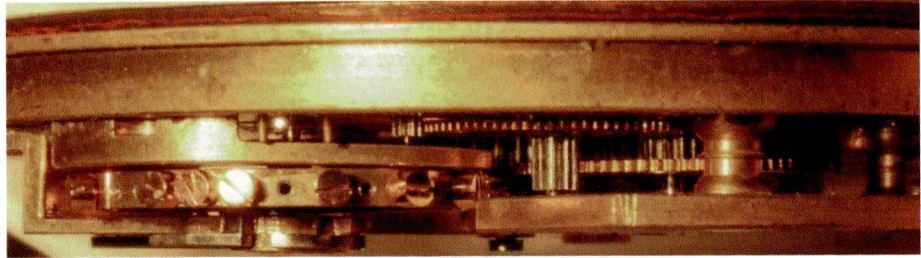

Fridlander watch movement 51965

More is written about both Alfred Fridlander, and Philip Cohen in the books: 'The Jews of Coventry' (Chapters III & IV) and in the 'Coventry Watchmakers Trail'.

See Chapter 10 'Coventry Synagogue' below to read of Alfred's leadership with Philip Cohen in establishing the Coventry Synagogue.

See Page 44, in the Watchmakers Detail Reference Section, for further details of his life and business including becoming Director, and then Chairman, of the Triumph Cycle Company.

As the Certified Secretary, A E Fridlander's signature appears on every Coventry Synagogue Marriage Register record at from 1871 to 1919

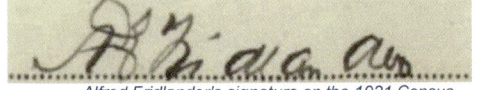

Alfred Fridlander's signature on the 1921 Census, where with the precision of a lifelong watchmaker he gives his age as 81 years & 7 Days

Fridlander watch movement signature

8. Commercial Travellers – Levin Joel, Jane Joel & Lewis Anidjah

Coventry's Jewish Watchmakers were significant, not only in establishing the Coventry Synagogue in 1870 (as we shall see in Chapter 10), but also in providing the Old Jewish Cemetery in 1863 (See Coventry Jewish Cemetery p.49 for more details). The early graves in the cemetery reflect the lives of many of the watchmakers from the earlier chapters. In investigating the other earliest graves, we find three not accounted for in those stories and we make the following encounters …

The fourth grave is that of Levin Joel. According to his gravestone, Levin Joel was a dear spirited man, the son of Yehuda Joel. Levin died on Sunday 7th Oct 1877 – the 1st of Tishri, 5638 in the Jewish calendar. He was buried two days later on Tuesday morning 3rd Tishri.

Buried next to him is David Anidjah who died in January the previous year.

We find the Joel family in Chapelfields, Coventry in 1861. Levin Joel, from Prussia; Jane Joel née Flatow born in Leith, Edinburgh; their children; and Jane's parents, Solomon & Maria Flatow who were also from Prussia.

Levin & Jane had been married in 1845 in Leeds. From there they moved to Manchester, arriving in Coventry around 1860.

Jane's mother, Maria, died in Coventry on Sunday 24th March 1861. There being no Jewish cemetery yet in Coventry, she was buried in the old Betholom Row Cemetery on what is now 5 Ways in the centre of Birmingham.

The 1861 census took place just 2 weeks later on the night of Sunday 7 April. The Joel family, still deeply mourning the death of Jane's mother, were resident at number 9 Duke Street in Coventry's new Chapelfields watchmaking district. Levin, aged 40, was a General Dealer, but soon to specialise in an area dear to our hearts …

Jane Joel was 32 years old. Levin & Jane have with them their five children: Lewis (8), Rosa (6), Augusta, who they called 'Gusta (4) and Henry, aged just 2½ months. A further daughter, Miriam, would be born two years later. Jane Levin's father, Solomon Flatau/Flatow, aged 69, and widowed just two weeks earlier, is staying with them.

The older children were born in Manchester. Henry was born in Coventry. This immigrant family, like so many, had moved around from place to place seeking work, the need to provide for their family, and to contribute to the ever-changing community around them.

Henry lived just 3 years before dying of scarlet fever – so easily treatable now. Rosa's life was, sadly, to be short lived too. She died in 1868 aged just 12 and was the first to be buried in the then new Jewish Cemetery, by the railway line on the London Road.

The family story unfolds more for us in the 1871 census. On the night of the census, 2nd April 1871, most of the family were in Coventry, but Levin was not …

The census shows the family living in Hearsall Terrace, on the Allesley Old Road in Chapelfields – the very heart of the Coventry's watch manufacturing area. Jane Joel is there with children Lewis, Augusta, Fanny & Miriam. There is then an unusual and delightful additional note indicating that her husband (Levin Joel) has 'gone to Leeds' and that he is a 'Watch Manufacturer employing 8 men and 4 boys'.

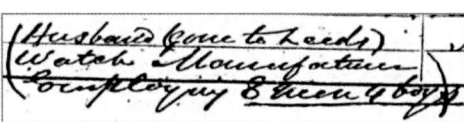

Note on Jane Joel's 1871 Census

So, we search in Leeds and find that on that Sunday night Levin Joel is lodging at the Inevelyan Hotel, near the railway station with a couple of dozen other commercial travellers. Levin shows his occupation as Watch Manufacturer. So, Levin well and truly, joins the ranks of Coventry's Jewish Watchmakers.

Even more beautifully woven into our story is that, among the commercial travellers at the hotel that night are Moss Fridlander Jnr, and Lewis Anidjah. Moss Fridlander Jnr. is the Birmingham Jewish Watchmaker, the son of David & Esther Fridlander, and the brother of Alfred Emmanuel Fridlander who stars throughout the story of Coventry's Jewish Watchmakers

Lewis Anidjah is the grandson of Coventry's David Anidjah. David is the second oldest grave in the Coventry Jewish Cemetery, and connects all the very early graves to the lives of the watchmakers

Jane Joel's brother was the famous art dealer Louis Victor Flatou. He paid a world record of £5,000 or more for William Powell Frith's oil painting, the Railway Station. He had a colourful life about which much could be written. He promised throughout his life to ensure that his sister would be provided for. However, when he died in 1867 at the age of just 47, the amount provided for Jane in his will had been recently reduced by 60%. Jane challenged in the Court of Probate for the full amount. Coventry Jewish watchmakers Philip Cohen and Jane's husband Levin Joel gave testimony, but the court found in favour of Louis's widow.

Levin & Jane's surviving four children, Augusta, Lewis, Fanny & Miriam were all married in the Coventry Synagogue between 1875 and 1884. Each of their marriage register entries bears the signature of Alfred Fridlander who was the Synagogue Secretary and Registrar.

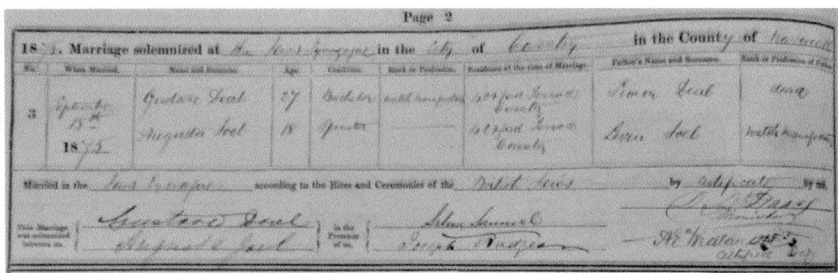

Coventry Synagogue Marriage Register of Augusta Joel & Gustave Deal 15 Sep 1875

Levin Joel died in 1877, leaving Jane with their surviving small children and the business. So, what became of his family and his business?

Jane Joel took over and continued the business, as reported in the 1881 census:

The family are now living at number 4 Oxford Terrace, just 3 doors down from Philip & Priscilla Cohen. It is now Jane who appears in her own right as a Watch Manufacturer – the first woman to do so. Her daughter, Fanny, aged 22, is still with her, as is her nephew Benjamin Flatow.

Jane was a very generous and trusting person. She had taken out a loan as a gift to her daughter on her wedding in the newly build Coventry Synagogue. She had also put her name as guarantees on loans taken out by her children.

Sadly, with the downturn of the watchmaking trade in Coventry, and with her children defaulting on their loans, Jane was served with a notice of bankruptcy on 1st December 1885.

Together with her son Lewis – who had himself been part of the cause of her financial difficulties - she sought new opportunities in South Africa. She lived there a further 11 years in East London. The Jewish Chronicle, reported on Friday 26th. November, 1896 that she had died there.

The Joel family story with its twists and turns, and its highlights and sorrows, weaves together the last strands of the Coventry Jewish Watchmakers stories as can be seen from the lives of those who were buried in the London Road Cemetery, which had been acquired through their endeavours.

9. A Jewish Wedding in Coventry - Lizzie Baum to Jacob Landau

Coventry Synagogue was the venue for the marriage of Lizzie Baum to Jacob Landau on 22nd July 1891. Lizzie was the daughter of Coventry Watchmaker Marks Baum. The bridegroom was a jeweller from Birmingham. News of the wedding was celebrated in the Coventry Standard newspaper as follows: https://www.newspapers.com/image/787443729/

JEWISH WEDDING IN COVENTRY

A Jewish marriage in this locality is so rare an occurrence as to entitle it to some notice.

No such ceremony has taken place before last Wednesday, for six or seven years. At three o'clock a congregation consisting of friends of the bride and bridegroom, and residents in the immediate neighbourhood, assembled at the synagogue, where the wedding took place.

It may be incidentally mentioned that a marriage cannot be solemnised unless ten men are present, that number being requisite to form a synagogue.

A Birmingham rabbi (Rev. G. Emmanuel, B.A.) officiated and was apparelled in a flowing velvet gown and a cap of the same material.

The bridegroom and his friends took their place beneath a canopy which was supported by four poles, and on a signal being given that all was in readiness, the bride came forth from an anteroom leaning on the arm of a female friend.

The marriage ceremony was exceedingly short. The happy man produced a ring, which the rabbi took and afterwards returned to him.

The latter then addressed the bride and groom, separately and together, in the English language, and very happily conceived the address was. Much sage council was given and he expressed his gratification on hearing that the union was one founded on love. The purpose of marriage was not to enable man and woman to live a pleasanter life, but that they might help one another.

A glass of wine was then handed to the rabbi, who pronounced a blessing over it, and it was next given to the bride and bridegroom, who both tasted it.

Afterwards the glass was put on the ground and broken by the husband to cries of "Mazel tov" (Good luck) from the bystanders, in token that, as the glass could never be put together again, so the marriage now made could not be broken.

The rabbi offered prayer in English, and the ceremony was over – except the kissing of the bride by many of her friends.

In the Jewish marriage service, there is no declaration by either the man or the woman of the one taking the other as wife and husband: that is assumed by their presence.

The service, which was followed with evident interest by the Gentile ladies present, was attended by a congregation of about 40 persons.

A heavy thunderstorm prevailed during the whole time the ceremony occurred.

10. Coventry Synagogue

Coventry Jewish Watchmakers were key in the moves that lead to the building of the Coventry Synagogue. The synagogue opened in 1870 in Barras Lane …

It was Alfred Emanuel Fridlander who led much of the planning and development work, submitting the plans for approval:

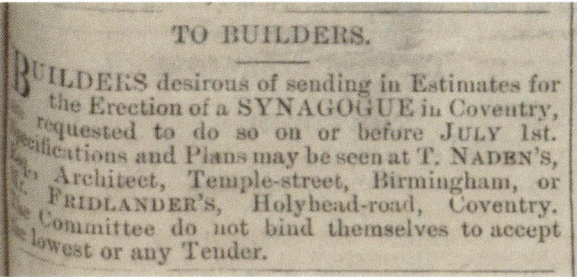

Coventry Standard 25 Jun 1869 - Builders Estimates for Synagogue Requested

Coventry Standard 3 Sep 1869 - Synagogue Plans Rejected

Coventry Standard 3 Sep 1869 - Synagogue Plans Passed

Coventry Standard 15 Apr 1870 Synagogue Surveyors Report Approved

The land for the synagogue was held in a charitable trust called The Philip Emanuel Cohen Charity.

All three of its trustees were watchmakers: **Philip Cohen**, **Francis Silveston** & **Abraham Emanuel**.

Extract from Synagogue Land Registry Deed

As you would expect, the Jewish Watchmakers played key roles in the celebration of the opening of the synagogue in 1870…

The dedication service was led by the Chief Rabbi, the Very Rev. Dr. Nathan Adler.

Following this, in the evening a grand banquet was held at St. Mary's Hall, at which Mr. **Philip Cohen** presided, being supported by the Chief Rabbi, … Mr, and Mrs. David, and many others.

Mr. **Joseph Radges** gave a toast to "The Architect and Builders".

At the conclusion of the banquet, **Alfred Emanuel Fridlander**, the honorary secretary of the congregation, was called to the chair, and said, that the first toast he had to propose was that of Her Majesty the Queen [Victoria], which was heartily responded to.

The Chief Rabbi noted that Mr. **Fridlander** had helped to establish a school at the synagogue.

Mr Levin had the privilege of proposing the toast to the health of the Chairman (**Alfred Emanuel Fridlander**).

He said, "if any one gentleman in particular had contributed to the bringing about the consummation of the building they had consecrated that day, it was Mr. **Fridlander**. He also spoke in eulogistic terms of the services rendered by their worthy chairman in the formation of the school. He had great pleasure proposing the health of Mr. and Mrs. **Fridlander**".

The Coventry Standard records that:
"Mr **Fridlander** acknowledged the compliment in suitable terms".

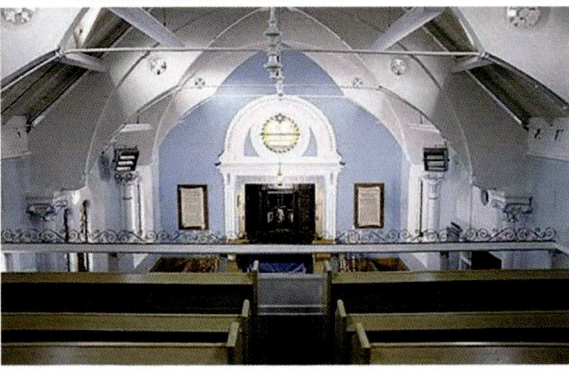

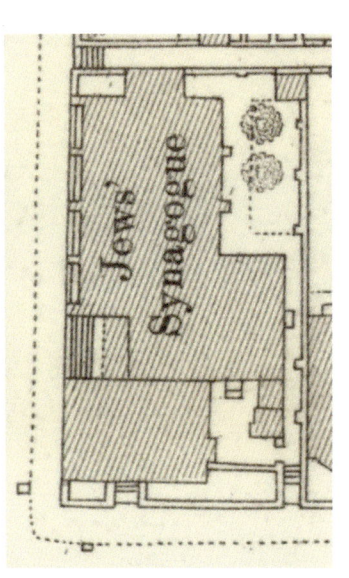

Coventry Synagogue from town plan of 1886. Note the garden area behind the synagogue at that time

Ordnance Survey 1:500 map surveyed 1886. Published 1888. Reproduced with the permission of the National Library of Scotland

11. Coventry Hebrew Congregation – An Astonishing Discovery

We initially found 7 Jewish Coventry watchmakers and were delighted to see their family trees intertwined through marriage.

The booklet of the Coventry Watchmakers Heritage Trail mentions **Mendel & Joseph Radges** among the Jewish Watchmakers. Mendel, his sons, Joseph & Augustus, and his grandsons Louis & Ernest were all in the watch making trade.

The 1970 book 'The Jews of Coventry', also references **Francis Silveston** as a Watch Manufacturer. The book records the wedding of their daughter Sarah Silveston as being the first marriage to take place in the new synagogue ('Shul' in Yiddish).

Cross referencing in 'The Jews of Coventry' book, we find that **Philip Cohen, Alfred Fridlander, Francis Silveston & Joseph Radges** were 4 of the first 5 Presidents of the Coventry Hebrew Congregation. Which prompted the now obvious question: who was the fifth President of the Congregation and what was his occupation ...

The fifth name is **Marks Baum** & his wife Caroline. They also were from Germany. They lived in Fleet Street, then Spon Street and then on the Holyhead Road, and, yes, you have guessed right, he also was a watch manufacturer - completing the set of the first five presidents of the Coventry Hebrew Congregation!

See the Watchmakers Family Trees in the next Chapter for all these connections.

We saw above the significant role of the Jewish Watchmakers contributing to the building of the Coventry Synagogue in 1870.

By the turn of the twentieth century the Coventry watch industry was in decline. Many had moved to London - including most of the Jewish watchmakers - as stiff competition grew from elsewhere. Coventry then diversified yet again becoming leading lights in bicycle, and then car manufacture.

The decline in the watch trade was mirrored in the Jewish community with a great reduction in membership. By 1902 the Shul was closed - though, it re-opened 4 years later as the community expanded with new families.

So, it is clear that at the time the Shul was not viable without the Watchmakers ... and it seems most likely, that the small Coventry Jewish community only grew to the size where the great task of building the Shul (Yiddish for synagogue) could be realised, because of the Watchmakers ... So, it seems reasonable to conclude that if there had been no Jewish Watchmakers then there would have been no Coventry Shul!

We are greatly indebted to their pioneering work, and their commitment to the community, for the Shul that is now here.

Watchmakers Family Trees

Not surprisingly there were many connections between the Jewish watchmaker families – both business partnerships and intermarriage.

Here are some of the connections we have found. The following family trees show the detail.

The following were all Coventry Jewish Watchmakers and/or silversmiths:

Philip Cohen:
- Philip Cohen's sister, Pauline Cohen, wife of **Francis Silveston** Frances & Pauline Silveston's daughter, Sarah's marriage, was the first in the new Coventry Synagogue on 29th Nov 1871
- Philip Cohen's nephew, **Selim Samuel**;
- Philip Cohen's brother-in-law, **Philip Solomon**;
- Philip Solomon's son-in-law, **Alfred Emanuel Fridlander**;

Alfred Emanuel Fridlander:
- Alfred's father, **David Fridlander** (Silversmith in Birmingham)
- Alfred's uncle, **Moss Fridlander Snr**;
- Alfred's brother, **Moss Fridlander Jnr**;
- Alfred's brother-in-law, **Michael Klean;**
- Alfred's brother-in-law, **Marks Baum**

Marks Baum:
- Their daughter, Lizzie's marriage, on 22nd July 1891 was reported in detail in the Coventry Standard
- Marks son, **Michael Baum**

Mendel Radges:
- Mendel's son, **Joseph Radges**
- Mendel's son, **Augustus Radges**
- Mendel's grandson, **Louis Henry Radges**
- Mendel's grandson, **Ernest Radges**

See detail Family Trees overleaf

*Alfred Fridlander Watch Movement.
Photo courtesy of Tony Barber.*

12. Coventry Jewish Watchmakers Family Tree Overview

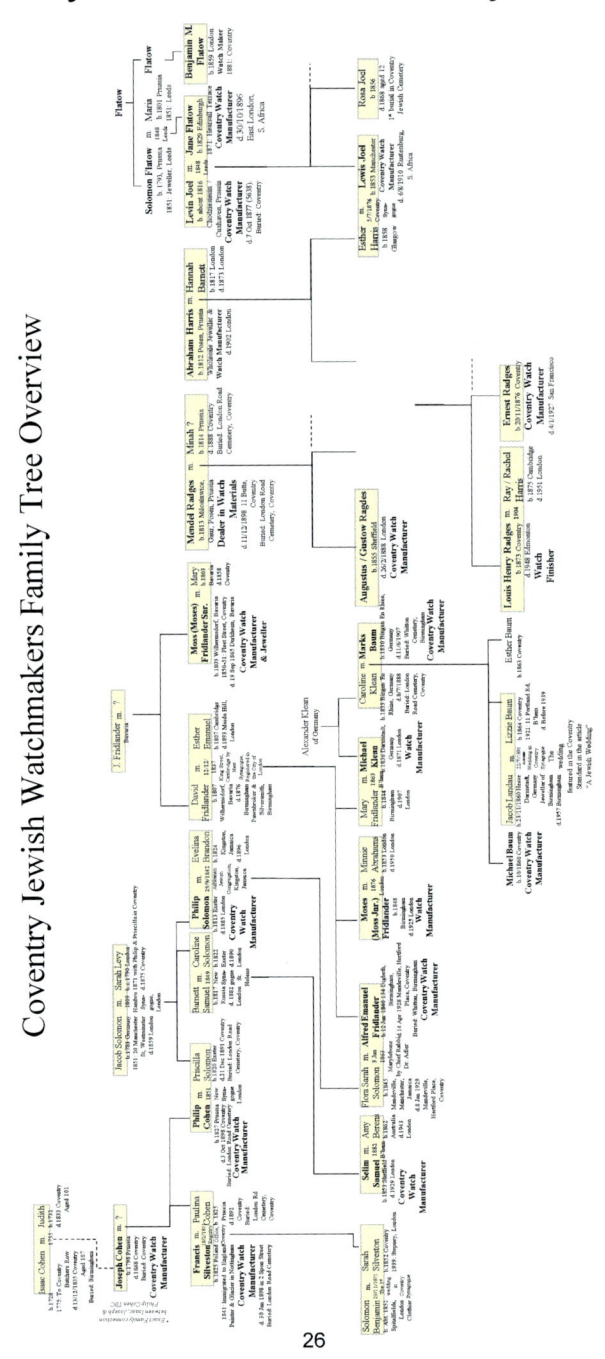

13. Cohen, Solomon, Samuel, Fridlander, Klean, Baum & Silveston Family Tree

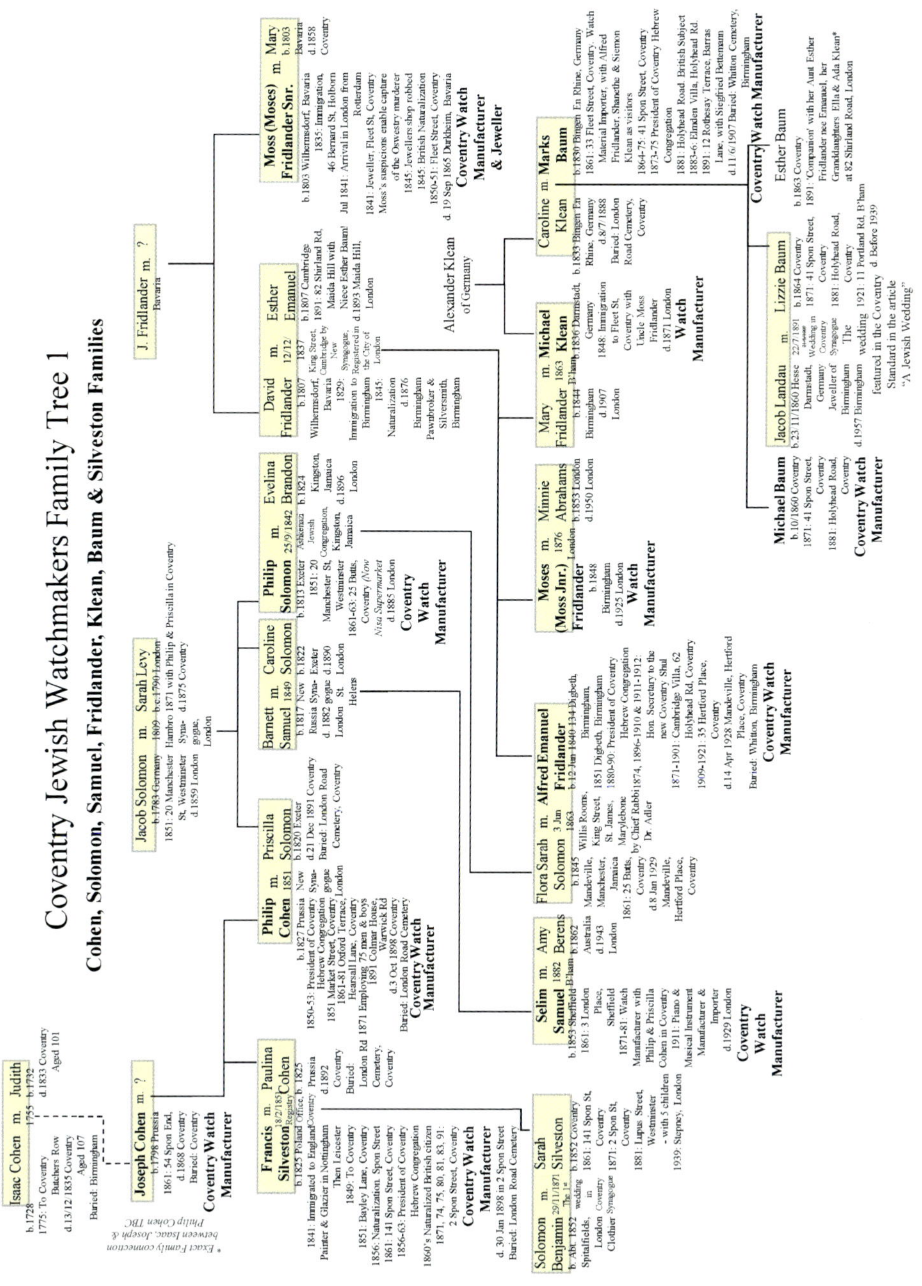
Coventry Jewish Watchmakers Family Tree 1
Cohen, Solomon, Samuel, Fridlander, Klean, Baum & Silveston Families
Isaac Cohen m. 1755 Judith
b.1728
1775: To Coventry Butchers Row
d.13/12/1835 Coventry Aged 107
Buried: Birmingham
b.1732
d.1833 Coventry Aged 101
* Exact Family connection between Isaac, Joseph & Philip Cohen TBC
Joseph Cohen m. ?
b.1798 Prussia
1861: 54 Spon End,
d.1868 Coventry
Buried: Coventry
Coventry Watch Manufacturer
Jacob Solomon m. 1809 Hambro Syna-gogue, London Sarah Levy
b.1783 Germany
1851: 20 Manchester St, Westminster
d.1859 London
b.c.1790 London
1871 with Philip & Priscilla in Coventry
d.1875 Coventry
J. Fridlander m. ?
Bavaria
Francis Silveston m. 18/2/1851 Registry Office, Coventry Paulina Cohen
b.1825 Poland
1841: Immigrated to England
Painter & Glazier in Nottingham
Then Leicester
1849: To Coventry
1851: Bayley Lane, Coventry
1856: Naturalization. Spon Street
1861: 141 Spon Street, Coventry
1856-63: President of Coventry Hebrew Congregation
1860's Naturalized British citizen
1871, 74, 75, 80, 81, 83, 91: 2 Spon Street, Coventry
Coventry Watch Manufacturer
d. 30 Jan 1898 in 2 Spon Street
Buried: London Road Cemetery
b. 1825 Prussia
d.1892 Coventry
Buried: London Rd Cemetery, Coventry
Philip Cohen m. 1851 New Syna-gogue London Priscilla Solomon
b.1827 Prussia
1850-53: President of Coventry Hebrew Congregation
1851 Market Street, Coventry
1861-81 Oxford Terrace, Hearsall Lane, Coventry
1871 Employing 75 men & boys
1891 Colmar House, Warwick Rd
d.3 Oct 1898 Coventry
Buried: London Road Cemetery
Coventry Watch Manufacturer
b.1820 Exeter
d.21 Dec 1891 Coventry
Buried: London Road Cemetery, Coventry
Barnett Samuel m. 1849 New Syna-gogue St. Helens Caroline Solomon
b.1817 Russia
d. 1882 London
b.1822 Exeter
d.1890 London
Philip Solomon m. 25/9/1842 Ashkenazi Jewish Congregation, Kingston, Jamaica Evelina Brandon
b.1813 Exeter
1851: 20 Manchester St, Westminster
1861-63: 25 Butts, Coventry (Now Nisa Supermarket
d.1885 London
Coventry Watch Manufacturer
b.1824 Kingston, Jamaica
d.1896 London
David Fridlander m. 12/12/1837 King Street, Cambridge by New Synagogue, Registered in the City of London Esther Emanuel
b.1807 Wilhermsdorf, Bavaria
1829: Immigration to Birmingham
1845: Naturalization
d.1876 Birmingham
Pawnbroker & Silversmith, Birmingham
b.1807 Cambridge
1891: 82 Shirland Rd, Maida Hill with Niece Esther Baum!
d.1893 Maida Hill, London
Moss (Moses) Fridlander Snr. m. Mary
b.1803 Wilhermsdorf, Bavaria
1835: Immigration, 46 Bernard St, Holborn
Jul 1841: Arrival in London from Rotterdam
1841: Jeweller, Fleet St, Coventry
Moss's suspicions enable capture of the Oswestry murderer
1845: Jewellers shop robbed
1845: British Naturalization
1850-51: Fleet Street, Coventry
d. 19 Sep 1865 Durkheim, Bavaria
Coventry Watch Manufacturer & Jeweller
b.1803 Bavaria
d.1858 Coventry
Alexander Klean of Germany
Solomon Benjamin m. 29/11/1871 The 1st wedding in Coventry Synagogue Sarah Silveston
b. Abt. 1852 Spitalfields, London
Clothier
b.1852 Coventry
1861: 141 Spon St, Coventry
1871: 2 Spon St, Coventry
1881: Lupus Street, Westminster - with 5 children
1939: Stepney, London
Selim Samuel m. 1882 B'ham Amy Berens
b.1853 Sheffield
1861: 3 London Place, Sheffield
1871-81: Watch Manufacturer with Philip & Priscilla Cohen in Coventry
1911: Piano & Musical Instrument Manufacturer & Importer
d.1929 London
Coventry Watch Manufacturer
b.1862 Australia
d.1943 London
Flora Sarah Solomon m. 3 Jun 1863 Willis Rooms, King Street, St. James, Marylebone by Chief Rabbi Dr. Adler Alfred Emanuel Fridlander
b.1845 Mandeville, Manchester, Jamaica
1861: 25 Butts, Coventry
d.8 Jan 1929 Mandeville, Hertford Place, Coventry
b.12 Jun 1840 134 Digbeth, Birmingham,
1851 Digbeth, Birmingham
1880-90: President of Coventry Hebrew Congregation
1874, 1896-1910 & 1911-1912: Hon. Secretary to the new Coventry Shul
1871-1901: Cambridge Villa, 62 Holyhead Rd, Coventry
1909-1921: 35 Hertford Place, Coventry
d.14 Apr 1928 Mandeville, Hertford Place, Coventry
Buried: Whitton, Birmingham
Coventry Watch Manufacturer
Moses (Moss Jnr.) Fridlander m. 1876 London Minnie Abrahams
b.1848 Birmingham
d.1925 London
Watch Manufacturer
b.1853 London
d.1950 London
Mary Fridlander m. 1863 B'ham Michael Klean
b.1844 Birmingham
d.1907 London
b.1836 Darmstadt, Germany
1848: Immigration to Fleet St, Coventry with Uncle Moss Fridlander
d.1871 London
Watch Manufacturer
Caroline Klean m. Marks Baum
b.1833 Bingen En Rhine, Germany
d.8/7/1888
Buried: London Road Cemetery, Coventry
b.1830 Bingen En Rhine, Germany
1861: 33 Fleet Street, Coventry. Watch Material Importer, with Alfred Fridlander, Shanethe & Siemon Klean as visitors
1864-75: 41 Spon Street, Coventry
1873-75 President of Coventry Hebrew Congregation
1881: Holyhead Road. British Subject
1883-6: Elmden Villa, Holyhead Rd.
1891: 12 Rothesay Terrace, Barras Lane, with Siegfried Bettemann
d.11/6/1907 Buried: Whitton Cemetery, Birmingham
Coventry Watch Manufacturer
Michael Baum
b.10/1860 Coventry
1871: 41 Spon Street, Coventry
1881: Holyhead Road, Coventry
Coventry Watch Manufacturer
Jacob Landau m. 22/7/1891 Wedding in Coventry Synagogue Lizzie Baum
b.23/11/1860 Hesse Darmstadt, Germany
Jeweller of Birmingham
d.1957 Birmingham
The wedding featured in the Coventry Standard in the article "A Jewish Wedding"
b.1864 Coventry
1871: 41 Spon Street, Coventry
1881: Holyhead Road, Coventry
1921: 11 Portland Rd, B'ham
d. Before 1939
Esther Baum
b.1863 Coventry
1891: 'Companion' with her Aunt Esther Fridlander nee Emanuel, her Granddaughters Ella & Ada Klean* at 82 Shirland Road, London

14. Radges, Harris & Joel Families

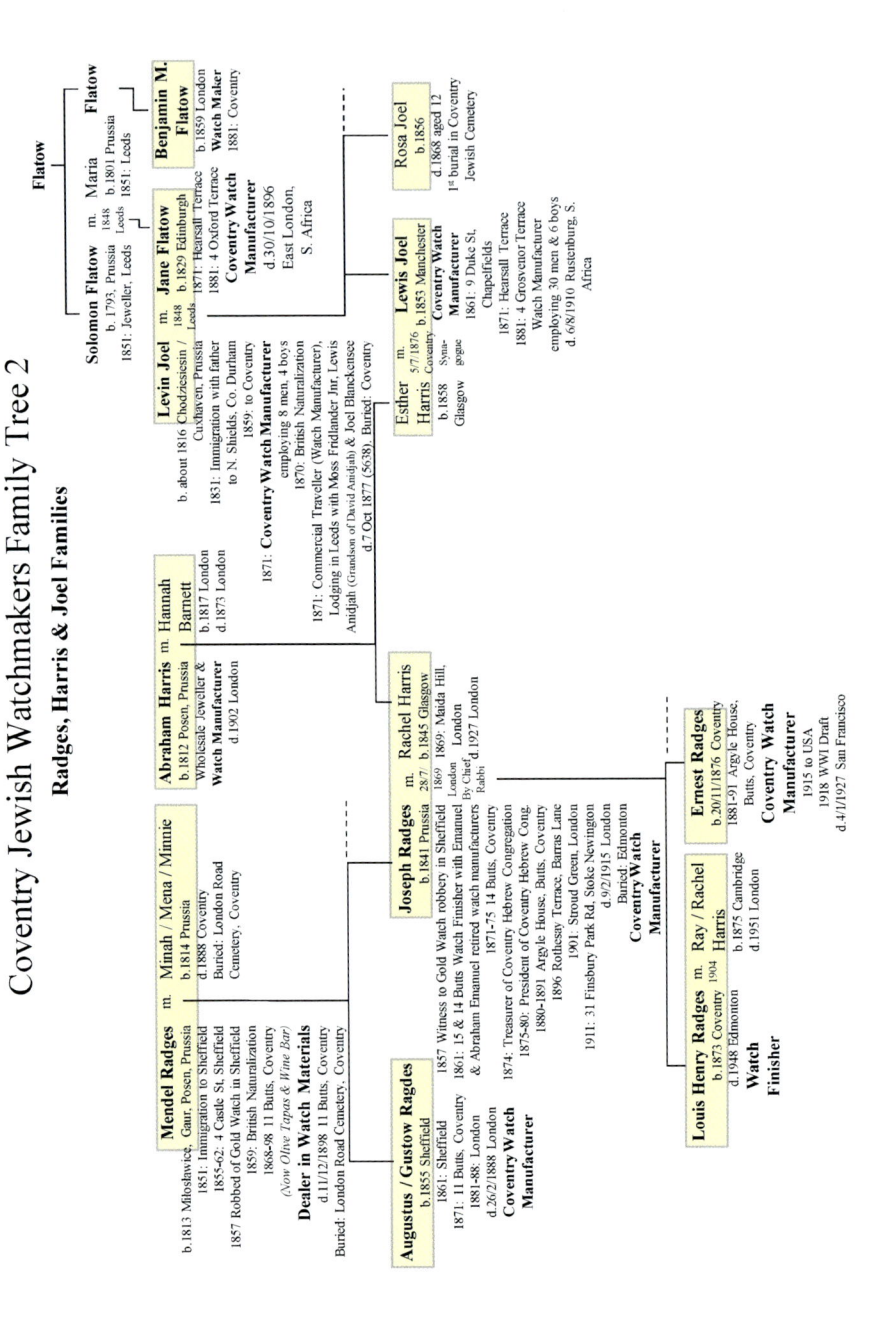

Coventry Jewish Watchmakers Family Tree 2
Radges, Harris & Joel Families

Flatow

Solomon Flatow m. 1848 **Maria** b.1793, Prussia / Leeds / b.1801 Prussia
1851: Jeweller, Leeds — 1851: Leeds

Benjamin M. Flatow
b.1859 London
Watch Maker
1881: Coventry

Jane Flatow b.1829 Edinburgh
1871: 4 Oxford Terrace
Coventry Watch Manufacturer
d.30/10/1896 East London, S. Africa

Levin Joel m. 1848 Jane Flatow / Leeds
b. about 1816 Chodziesiesin / Cuxhaven, Prussia
1831: Immigration with father to N. Shields, Co. Durham
1859: to Coventry
1871: **Coventry Watch Manufacturer** employing 8 men, 4 boys
1870: British Naturalization
1871: Commercial Traveller (Watch Manufacturer), Lodging in Leeds with Moss Fridlander Jnr, Lewis Anidjah (Grandson of David Anidjah) & Joel Blanckensee
d. 7 Oct 1877 (5638). Buried: Coventry

Rosa Joel b.1856
d.1868 aged 12
1st burial in Coventry Jewish Cemetery

Lewis Joel b.1855 Manchester
Coventry Watch Manufacturer
1861: 9 Duke St, Chapelfields
1871: Hearsall Terrace
1881: 4 Grosvenor Terrace Watch Manufacturer employing 30 men & 6 boys
d. 6/8/1910 Rustenburg, S. Africa

Esther Harris b.1858 Glasgow
m. 5/7/1876 Coventry Synagogue

Mendel Radges m. **Minah / Mena / Minnie**
b.1813 Miloszwiecz, Gaur, Posen, Prussia / b.1814 Prussia
1851: Immigration to Sheffield
1855-62: 4 Castle St, Sheffield
1857 Robbed of Gold Watch in Sheffield
1859: British Naturalization
1868-98 11 Butts, Coventry
Dealer in Watch Materials
(Now Olive Tapas & Wine Bar)
d.11/12/1898 11 Butts, Coventry
Buried: London Road Cemetery, Coventry

d.1888 Coventry
Buried: London Road Cemetery, Coventry

Abraham Harris m. **Hannah Barnett**
b.1812 Posen, Prussia / b.1817 London
Wholesale Jeweller & Watch Manufacturer
d.1902 London / d.1873 London

Augustus / Gustow Ragdes
b.1855 Sheffield
1871: 11 Butts, Coventry
1881-88: London
d.26/2/1888 London
Coventry Watch Manufacturer

Joseph Radges m. **Rachel Harris**
b.1841 Prussia / 28.7 / 1869: Maida Hill, London
1869 / b.1845 Coventry By Chief Rabbi
1857 Witness to Gold Watch robbery in Sheffield
1861: 15 & 14 Butts Watch Finisher with Emanuel & Abraham Emanuel retired watch manufacturers
1871-75 14 Butts, Coventry
1874: Treasurer of Coventry Hebrew Congregation
1875-80: President of Coventry Hebrew Cong.
1880-1891 Argyle House, Butts, Coventry
1896 Rothesay Terrace, Barras Lane
1901: Stroud Green, London
d.9/2/1915 London
1911: 31 Finsbury Park Rd, Stoke Newington
Buried: Edmonton
Coventry Watch Manufacturer

Louis Henry Radges m. 1904 **Ray / Rachel Harris**
b.1873 Coventry / b.1875 Cambridge
d.1948 Edmonton / d.1951 London
Watch Finisher

Ernest Radges
b.20/11/1876 Coventry
1881-91 Argyle House, Butts, Coventry
Coventry Watch Manufacturer
1915 to USA
1918 WWI Draft
d.4/1/1927 San Francisco

Where they Lived & Worked

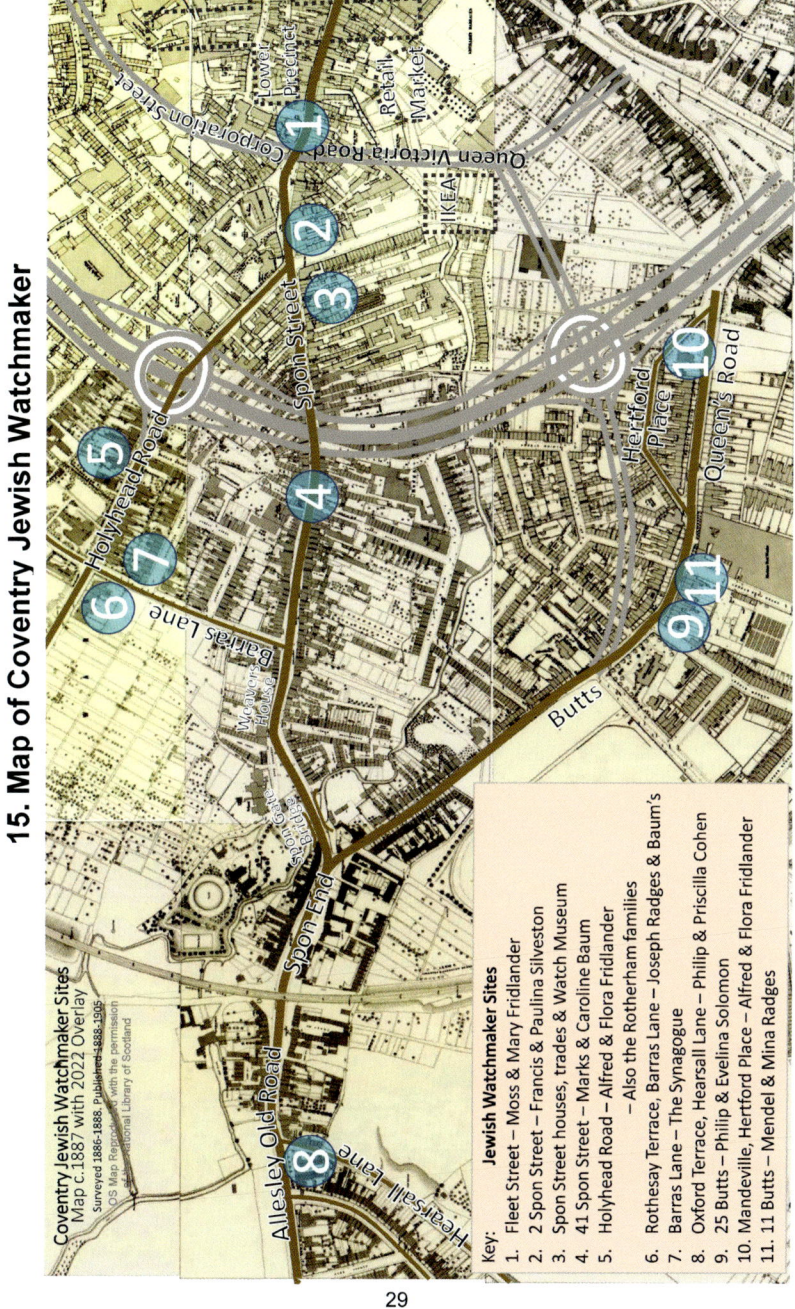

15. Map of Coventry Jewish Watchmaker

Coventry Jewish Watchmaker Sites
Map c.1887 with 2022 Overlay
Surveyed 1886-1888, Published 1888-1905
OS Map Reproduced with the permission of the National Library of Scotland

Key: Jewish Watchmaker Sites
1. Fleet Street – Moss & Mary Fridlander
2. 2 Spon Street – Francis & Paulina Silveston
3. Spon Street houses, trades & Watch Museum
4. 41 Spon Street – Marks & Caroline Baum
5. Holyhead Road – Alfred & Flora Fridlander
   – Also the Rotherham families
6. Rothesay Terrace, Barras Lane – Joseph Radges & Baum's
7. Barras Lane – The Synagogue
8. Oxford Terrace, Hearsall Lane – Philip & Priscilla Cohen
9. 25 Butts – Philip & Evelina Solomon
10. Mandeville, Hertford Place – Alfred & Flora Fridlander
11. 11 Butts – Mendel & Mina Radges

29

16. Cambridge Villa, Holyhead Road - Alfred & Flora Fridlander 1861-1901

We know from the Censuses that Alfred & Flora Fridlander lived in Cambridge Villa, Holyhead Road from 1861 until 1901. Cambridge Villa is no longer there, and its location was a mystery, as the house number is not given in any of the censuses. Reading the censuses for the Fridlanders and their neighbours in the order of the census records we find:

	1871	1881	1891	1901
		Holyhead Rd Alfred Kirby Watch Manufacturer	**Carlton House,** Holyhead Rd James Richardson	John Thomas Cycle Manufacturer
		Holyhead Rd James Richardson, Watch Manufacturer	Holyhead Rd Eliza Rotherham	80 Holyhead Rd
	Holyhead Rd John & Eliza Rotherham	Holyhead Rd Eliza Rotherham	Holyhead Rd	76 Holyhead Rd
	Holyhead Rd Charlotte Rotherham	**College House School**	**College House School**	**College House**
Fridlanders:	**Cambridge Villa**	Holyhead Rd	**Cambridge Villa**	Holyhead Rd
	School House	Holyhead Rd	Hampton Villa	60 Holyhead Rd
	Holyhead House	Holyhead Rd	Lund [?] Villa	58 Holyhead Rd
	186 Spon Street	Holyhead Rd	Eton Villa	56 Holyhead Rd

First, we see that in 1871 Alfred & Flora lived next door to Coventry's largest watchmakers: John Rotherham and his mother Charlotte Rotherham. Their factory in Spon Street employed over 500 people.

Compiling the large-scale Ordnance Survey maps surveyed between 1886 and 1888 we find Holyhead Road and Spon Street – the centre of the watchmaking trade as follows, with College House and Carlton house being helpfully labelled ...

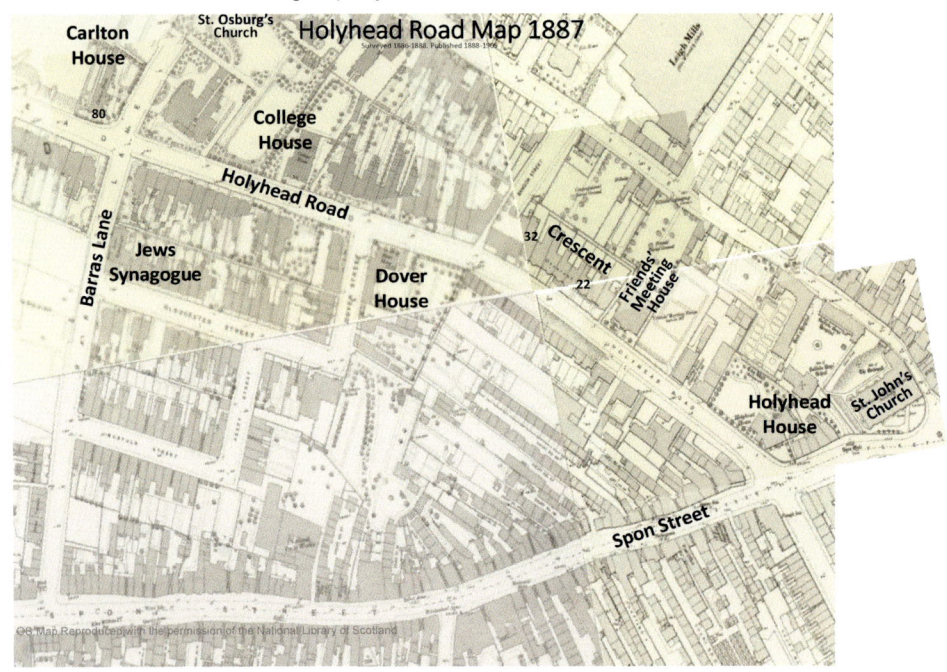

Holyhead Road Map 1887

From the census information we deduce that Alfred and Flora lived adjacent to the College House, with the two Rotherham family houses between them and Carlton House.

Overlaying the current Coventry Ring Road, which splits both the Holyhead Road and Spon Street, in half, this points us to the following location for the Fridlander and Rotherham families.

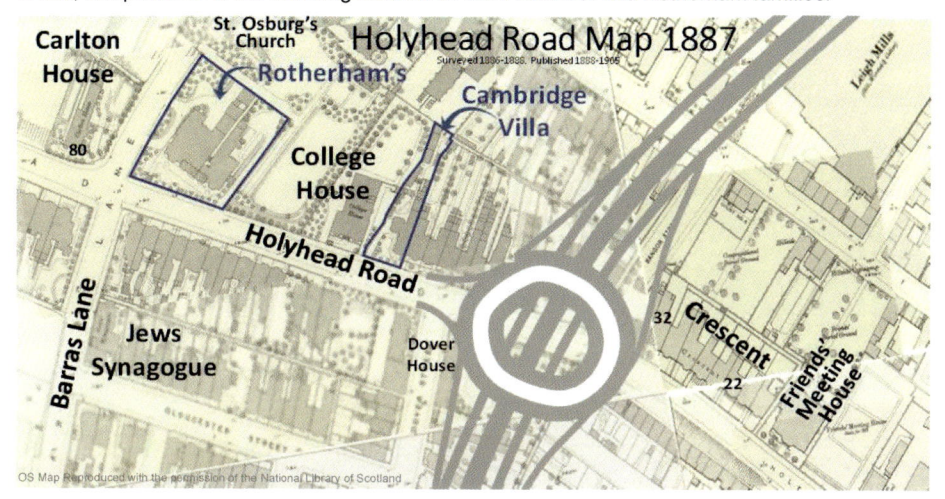

Carlton House is now replaced with the big ugly hotel on the on the corner of Holyhead Road and Barras Lane. College House is replaced by the school grounds of St Osburg's school and the Rotherhams' and Fridlanders' homes are no more.

Looking up Holyhead Road from the Ring Road towards Barras Lane. The Fridlanders' house would have been half way up on the right, with the Rotherhams' homes beyond them.

Here is the house and garden layout of Alfred & Flora's very modest semi-detached home. In keeping with the practice of most of the watch manufacturers, the many outbuildings would have been workshops for the business.

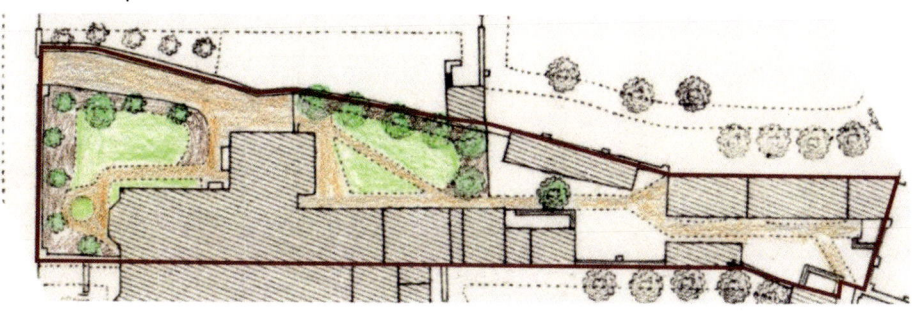

17. Mandeville, Hertford Place - Alfred & Flora Fridlander 1911-29

Later, when Alfred was a Councillor and a Justice of the Peace, they moved to the beautiful home which they called 'Mandeville' after the town where Flora was born in Jamaica. Mandeville was at the end of Hertford Place – now right up against the Ring Road by Queens Road Baptist Church.

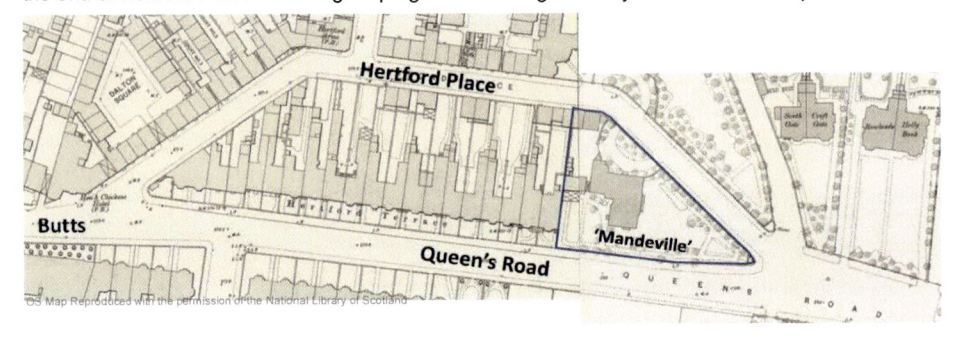

On the site now are Kings Chambers building …

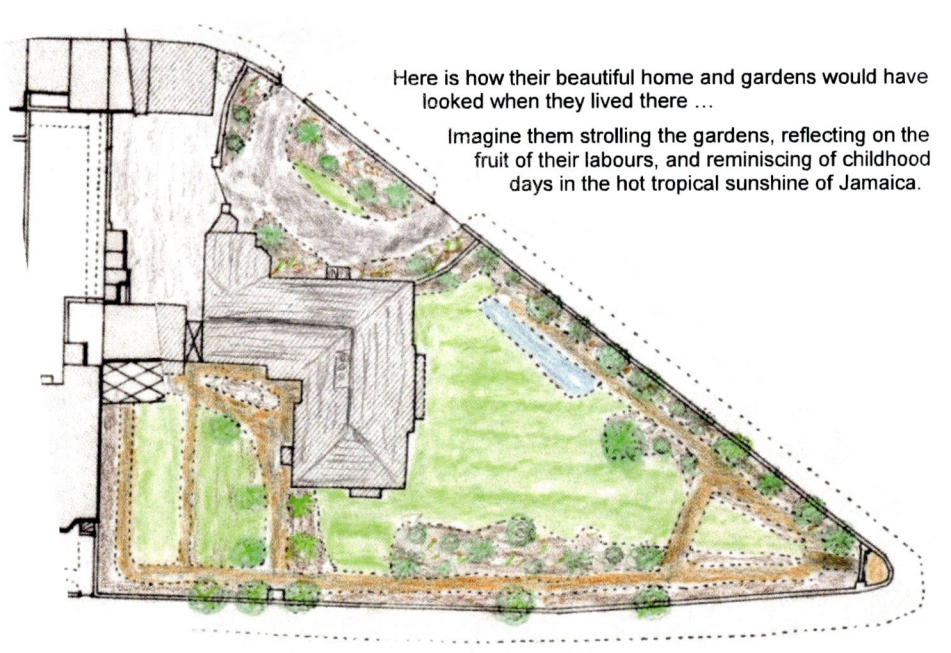

Here is how their beautiful home and gardens would have looked when they lived there …

Imagine them strolling the gardens, reflecting on the fruit of their labours, and reminiscing of childhood days in the hot tropical sunshine of Jamaica.

18. Rothesay Terrace, Barras Lane – Marks Baum 1891

Marks Baum's family lived in the splendour of the huge and ornate Rothsay Terrace houses in Barras Lane, directly opposite the newly built synagogue. Then numbered 12 Rothesay Terrace, tracing the route of the 1891 census enumerator leads us to find this is now number 49 Barras Lane. Here in 1891, we find the widowed Marks Baum with his daughters Jennie & Lizzie – who was shortly to be married (See Chapter 0). Also with them were the entrepreneurs Siegfried Bettmann, and his business partner in the Triumph Cycle Company, Moritz Shulte. Siegfried would go on to become Mayor of Coventry in 1913.

12 Rothesay Terrace - Now 49 Barras Lane

19. Bayley Lane – Francis & Paulina Silveston 1851

Francis Silveston was staying somewhere in Bayley Lane in 1851, with his bride Paulina Cohen.

They most likely posted their letters in this Victorian letter box on the old County Court building on the corner of Bayley Lane and Cuckoo Lane.

20. Butcher Row – Isaac & Judy Cohen 1722

Isaac and Judy Cohen hosted the very first Jewish worship in Coventry, in their home in a narrow alley way off the now destroyed Butchers Row in the City centre.

21. Fleet Street - Moss and Mary Fridlander 1841-58

Fleet Street was the location of the home and shop of Coventry's first Jewish Jeweller & Watchmaker, Moss and Mary Fridlander. It was here that Moss' cunning collaboration with the Coventry Chief of Police apprehended the Oswestry murderer.

Fleet Street, Coventry looking East from St John's church towards what is now the Lower Precinct

Fleet Street now replaced by Queen Victoria Street. No original buildings remain except for St. John's Church

We have no house number on Fleet Street for Moss & Mary Fridlander, and there are none on the maps of the time – just 3 pubs: Searching the censuses for Moss & Mary Fridlander and their neighbours does not pinpoint their location somewhere along Fleet Street.

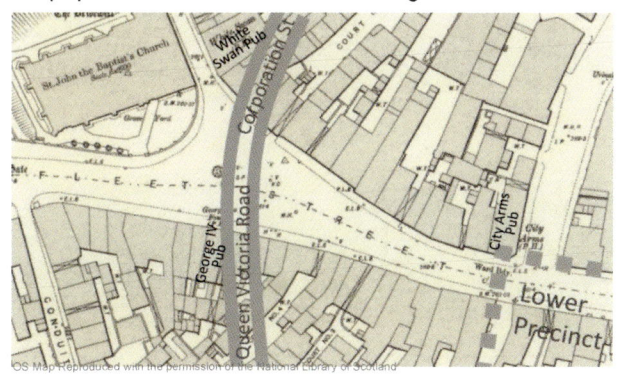

22. Oxford Terrace, Hearsall Lane – Philip Cohen 1861-81

On the corner of the Allesley Old Road and Hearsall Lane are the home and workshops of Philip & Priscilla Cohen's watch factory (now Matthew Lewis Displays). Philip & Priscilla's home – complete with Watchmakers Trail Blue Plaque – faces onto the Allesley Old Road, with the workshops and Top Shop of the factory behind. It is one of the highlights of preserved buildings in Coventry's Heritage Watchmakers Trail.

The Joel family also lived along Oxford Terrace.

The Assay Office Hallmark for Philip Cohen's silver watch cases made at Oxford Terrace

Watchmaker Top Shop behind the Cohen residence

Top Shops behind the adjacent Hills Watch Factory keeping their original form

23. Spon Street - Marks Baum 1871-75 & Francis Silveston 1861-98

Spon Street was the very heart of the watchmaking trade. Many of the factories were there, including the largest – Rotherham's. For centuries, Spon Street had been the main thoroughfare through Coventry for all business passing from London to the great ancient cities of Lichfield and Chester. The equivalent of the M6, A45 roads and the West Cost Mainline railway all combined together!

Spon Street looking towards St. Johns Church and Fleet Street. From the 1891 directory 'Where to buy in Coventry'

Now bisected by the Ring Road, what remains of lower Spon Street contains many medieval wooden framed buildings, both those originally located there, and those moved to be preserved from elsewhere in the City.

Thus, strolling down Spon Street you can still sense the hustle and bustle of times past, including some of the public houses within which much business must have been forged.

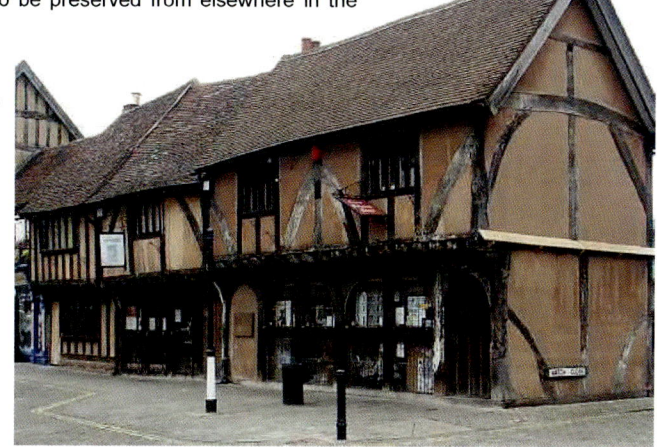

Nothing captures this better than The Coventry Watch Museum, located in the last remaining Spon Street Courtyard, and full of Coventry's watchmaking history treasures.

Coventry Watch Museum - The author's family with Leonard Heanes.

The Spon Street residences and workshops of Marks Baum at No. 41; and Francis Silveston at No.141 and, later, No. 2 are, alas, no more.

Display boards opposite the Coventry Watch Museum show the many trades which have taken place in Spon Street over the years including watchmaking.

WATCHMAKER

The display was produced Kenning Illustration & Creative Design for Coventry City Council.
Permission kindly granted to share this here.
The watch maker in the illustrations is Tony Barber of Coventry Watch Museum.

24. The Butts – Radges Family 1868-98 & Solomons 1861

For those living in Coventry, the Butts is well known as the road where what was the Technical College was adjacent to the Ring Road. Whilst Butts Road now curves around to join the Ring Road, it used to go straight on, and, indeed, there is still a residual part of the Butts which continues through to what is now (2022) the Ramada Hotel opposite a little row of small shops and eateries. Here we find the homes of some of our Jewish Watchmakers …

For 30 years from 1868, Number 11 – now the Olive Tapas & Wine Bar - was the home of Mendel & Minah **Radges**. Their children, Joseph & Augustus Radges, and grandchildren, Louis & Ernest Radges – each of them Watch Manufacturers – all started their lives here.

At Number 25 Butts – now the Nisa Local supermarket – **Philip Solomon** & Evelina Solomon née Brandon resided in 1861 – with their daughter Flora who would, shortly after, marry **Alfred Emanuel Fridlander**.

Notice the old painted BICYCLE sign which still tells of the next industry that would flood Coventry after the watch trade, and into which **Alfred Fridlander** would diversify at the Triumph Cycle Company.

Watchmakers Detail Reference

This Chapter provides a detailed catalogue of the records and events, many of which are expounded in the stories of the earlier chapters:

Contents

Isaac & Judy Cohen	41
David & Esther Fridlander	42
Moss & Mary Fridlander Snr (Brother of David Fridlander)	43
Alfred Emanuel & Flora Fridlander	44
Mary Emanuel Fridlander & Michael Klean	45
Philip, Evelina & Flora Solomon	46
Philip Cohen & Priscilla Solomon	46
Mendel, Minah & Joseph Radges	47
Francis & Paulina Silveston	47
Marks Baum	48
Levin & Jane Joel	48

Isaac & Judy Cohen

1728 – Isaac Cohen born, presumably in Prussia, as he served in the Prussian army.

1732 – Judy Cohen born. Maiden name not known to us.

1740-1763 – During at least some of this period Isaac served as Sergeant Major in the Prussian Army during the Wars of Frederick the Great. Frederick lived 1712 – 1786 and was King in Prussia from 1740 until 1786. The conflicts in which Isaac Cohen is likely to have served were the 1st Silesian War 1740-42, the 2nd Silesian War 1744-45, the 3rd Silesian War & 7 Years War 1756-63.

1775 – Isaac & Judy Cohen move to Coventry. According to Benjamin Poole's 1852 history of Coventry, Isaac lived in an obscure little tenement on the north side of a passage leading from Great Butcher to Priory Row.

1798 – Joseph Cohen born in Prussia – maybe a relative of Isaac and/or Philip Cohen?

1826 – Philip Cohen born in Prussia, supposedly to Isaac (died aged 99) & Judy Cohen (died aged 95)

1832 – Isaac & Judy's daughter "Fair Rosamund" performs as Britannia in the celebration procession for the passing of the Reform Act

1833 – Judy Cohen died aged 101 after 78 years of marriage

1835 – Isaac Cohen died aged 107

1868 – Joseph Cohen dies and is buried in Coventry Jewish Cemetery. His gravestone is inscribed: An affectionate son raises this monument a pious tribute to the memory of an honoured father. Given Philip Cohen's role in the community at the time, he is likely the son referenced in this touching tribute.

David & Esther Fridlander

1807 – David Fridlander was born about 1807 Wilhermsdorf, Bavaria

1828/9 – David Fridlander immigration to Birmingham

1828 – Fridlander & Davies Britannia Metal Manufacturers Partnership dissolved in Birmingham. Possibly David Fridlander?

1837 – David Fridlander married Esther Emanuel at her family home in King Street, Cambridge under the auspices of the New Synagogue of German & Polish Jews. It was registered in the City of London. Ether Emanuel was born about 1807 in Cambridge

1841 – David & Esther Fridlander, pawnbroker, age c.30 with Alfred age 1, at 134 Digbeth, Birmingham

1845 – British Naturalization

1846 – Death of Frederic Fridlander aged 4, son of David & Esther Fridlander, pawnbroker of Digbeth

1851 – David & Esther Fridlander, pawnbroker, age 43 with Alfred (10), Mary (7), Moses (3) [later also known as Moss] & Augustus (2). Children all born in Warwickshire [i.e. Birmingham]. David born in Bavaria & Esther born in Cambridgeshire

1861 – David & Esther E Fridlander, 53, trade now silversmith, 376 Bristol Rd, Birmingham, with children Mary E (17) & Moses (12) scholars
[No record found yet for Alfred or Augustus or Adelaide?]

1861 – David & Moss Fridlander's watch manufacturing partnership with Michael Klean dissolved

1871 – David (64) & Esther (64) Fridlander still living at 376 Bristol Rd, Edgbaston
with Augustus (22) Jeweller, Adelaide (20),
and granddaughter Ella Klean (7) [Daughter of Mary Fridlander & Michael Klean].

1876 – David died Jan 1876 in King's Norton, Worcestershire

1881 – Esther (73) widow now living with son, Moss Fridlander (33) watch manufacturer,
46 Bernard St, Finsbury. [This Moss (Moses) Fridlander was the nephew of Moss Fridlander (Coventry) below], and his wife Minnie (28), and their children (Katherine (4) & Mabel (2) in 46 Bernard Street, Finsbury, together with Ellen Klean (16) niece of Moss (Moses) Fridlander. Ellen is the daughter of Mary Fridlander & Michael Klean.

1891 – Esther (83) with her granddaughters, Ella (27) & Ada (24) Klean, and niece Esther Baum (31) at 82 Shirland Road, Paddington. Family connection to Esther Baum thought to be via daughter Mary Fridlander's marriage to Michael Klean, whose sister Caroline Klean married Marks Baum – the parents of Esther Baum!.

1893 – Esther died 20 Jan 1893. She lived at 82 Shirland Road, Maida Hill, London.
She left her effects to her sons Alfred Emanuel Fridlander, watch manufacturer & Moss Fridlander jeweller. Effects valued at £1,310 9s 2d.

Moss & Mary Fridlander Snr (Brother of David Fridlander)

1803 – Moss Fridlander born in Wilhermsdorf, Bavaria

1835 – Moss Fridlander Immigration to 46 Bernard St, Holborn

1841 – Moss Fridlander Arrival in London from Rotterdam

1841 – Moss Fridlander, traveller, age c.35, with Joseph Cohen, hawker, Fleet Street, Coventry

1841 – The suspicions of Moss Fridlander's, watchmaker and silversmith, residing in Fleet-street, Coventry, lead to the capture of the Oswestry murderer and his accomplice.

1845 – George Smith aged 19, committed on charge of feloniously breaking and entering Moss & Mary Fridlander's jeweller's shop

1845 – Moss Fridlander British Naturalization

1850 – Business Directory shows Moss Fri(e)dlander, watch manufacturer in Fleet Street

1851 – Moss Fridlander, & Mary, both age 48 in Fleet Street, Coventry.
Watchmaker Employs 9 men. British Subject born in Germany, Bavaria.
With Michael Klean (Klerin), nephew, 16, Watchmaker Apprentice
[Michael Klean, born Darmstadt, Hessen, Germany would in 1863 marry Mary Fridlander the daughter of David & Esther Fridlander]

1852 – The Conservatives objected to Moss Fridlander being included in the revised voters list on account of him being an alien. The presentation of his naturalisation papers was also challenged, but later accepted. 'Abraham Emmanuel, a Jew' was also challenged, but managed to prove that he was born in England

1857 – Mr. Fridlander [Moss or Alfred?] given notice by The Inspector of Nuisances in his report to the local Board of Health of the City of Coventry to erect a water-closet upon his property, situated in Victoria Street, Coventry

1858 – Mary Fridlander dies in Coventry

1861 – No census record found for Moss Fridlander

1861 – David & Moss Fridlander's watch manufacturing partnership with Michael Klean dissolved

1865 – Moss Fridlander, formerly of Coventry, late of 10 South Street Finsbury, widower, died 19 Sep 1865 at Durkheim, Bavaria. Probate to David Fridlander of Kensington Place, Bristol Road, Birmingham, brother of the deceased. Effects valued at under £5,000. [i.e. over £4,000]

Alfred Emanuel & Flora Fridlander

1840 – Alfred Emanuel Fridlander born 12 June 1840 to David Fridlander, a pawnbroker, & Esther Fridlander née Emanuel at 134 Digbeth, Birmingham

1841 & 1851 – Alfred with David & Esther Fridlander in Digbeth, Birmingham

1853-56 & 1880-90 – Alfred Fridlander President of the Coventry Hebrew Congregation

1861 – Alfred Fridlander, a visitor with Marks & Caroline Baum, Watch Material Dealer from Germany.
Marks & Caroline's children Esther (2) & Michael (5 months) born in Coventry.
David Fridlander age 20, born in Birmingham. Occupation: watch maker.
Other visitors Simeon Klean (20) & Shanethe (?) Klean (21), watch maker born in Germany.
[Simeon son of Alexander & Elsie Klean/Klein, naturalised 1914,
probably the brother of Michael Klean above. Shanethe possibly Leonora Defries?]

1863 – Alfred Emanuel Fridlander marries Flora Sarah Solomon on 3rd June 1863 by Chief Rabbi Dr. Nathan Adler in the Congregation of the New Synagogue of German & Polish Jews at the Willis Rooms, King Street, St. James, Marylebone, London. Flora, age 18, of 25 Butts, Coventry, was the daughter of Philip Solomon, watch manufacturer.

> Willis's Rooms had a 'large-ball room about one hundred feet in length, by forty feet in width; it is chastely decorated with gilt columns and pilasters, classic medallions, mirrors, &c., and is lit with gas in cut-glass lustres'.
> The Morning Advertiser, Monday 8 June 1863: MARRIAGES. On the 3rd inst, at Willis Rooms, St, James's, by the Rev. Dr. Adler, assisted by the Rev. A. Barnett, Mr. Alfred Fridlander, of Holyhead-road, Coventry, eldest son David Fridlander, Esq., of Birmingham, to Flora, only child of Mr. Philip Solomon, Coventry, formerly of Jamaica.

> On the 3rd inst., at Willis's Rooms, St. James's, London, by the Rev. Dr. Adler, assisted by the Rev. A. Barnett, Mr. Alfred Fridlander, of Hollyhead-road, Coventry, eldest son of David Fridlander, Esq., of Birmingham, to Flora, only child of Mr. Philip Solomon, of Coventry, formerly of Jamaica.

1863 – Alfred Fridlander Hon Sec to Coventry Volunteer Fire Brigade

1863-1866 – Alfred Fridlander, watch manufacturer of Holyhead Road elected in local elections for White Friars Ward with James Marriott.

1864 – Alfred Emanuel Fridlander offered a silver watch value £4 10s. as first prize in the second annual Swimming Matches at Coventry Baths, Hales Street

1864 – Alfred Emanuel Fridlander advocates Coventry Council for free public library

1864 – Alfred Emanuel Fridlander, manufacturer of Coventry initiated into the Birmingham 'Faithful Lodge' of the Freemasons on 8 Nov 1864

1869 – Alfred Fridlander's plans for a synagogue in Barras Lane were on 3 Sep 1869 rejected and referred to the General Works Committee. The plans were passed by the GWC on 17 Sep 1869

1871 – Alfred (30) & Flora Sarah Fridlander (26) born in Jamaica,
with children Annie Esther (7), Adelaide Rachel (5) & Ernest D E (5 months)
plus 2 servants at Cambridge Villa, Holyhead Road, Coventry.
Watch manufacturer employing 30 men and 6 boys.

1874-1912 – Alfred Fridlander was the Honorary Secretary of the newly built Coventry Synagogue.

1880-90 – Alfred Fridlander President of Coventry Hebrew Congregation

1881 – Alfred (40) & Flora (36) Watch Manufacturer, Holyhead Road with niece Ada Klean (14)

1887 – A Fridlander watch was 25th in the prize list of the Royal Society Kew Observatory Watch Ratings.

1888 – Fridlander watch is 13th in the Royal Society Kew Observatory Watch Ratings.

1889 – Fridlander watch achieves the first of many top 1st place awards for the most accurate watch in the country.

1891 – Alfred (50) & Flora (46) watch manufacturer, Cambridge Villa, Holyhead Road
with Annie (27), Ada (25) & Ernest (20)

1897 – Director of Triumph Cycle Co, Auto Machinery Co and Leigh Mills Co when the New Triumph Cycle Co, of which he was also a director, was opened for public subscription

1900 – Fridlander 'keyless going-barrel Karrusel lever watch' No. 25,582, obtained 90.1 marks out of a maximum of 100. This is the first English lever watch to reach the 90 marks limit, and its performance is the best since 1892 – The National Physical Laboratory, (Royal Society, Kew Observatory) Watch Ratings

1901 – Alfred (60) & Flora (56). Watch Manufacturer (J.P.) on Holyhead Road,
with Annie (36) & Ernest (30)

1904 – Chairman of Triumph Cycle Co

1909 – Alfred Fridlander, watch manufacturer in Hertford Place

1911 – Alfred (70) & Flora (66). Retired watch manufacturer, Hertford Place, back of Hertford Terrace, Coventry, with Annie (47) single; Ernest (40) & Ethel (37) Painter Artists married 1905.
Alfred and Flora had been married for 47 years. They had 4 children, 3 of whom were still alive. The census record is signed by Alfred.

1921 – Alfred (81) & Flora (80) 35 Herford Place, Coventry

1928 – Alfred Fridlander of Mandeville, Hertford Place, Coventry died 14 Apr 1928 leaving an estate of £85,384 3s. 6d. to his widow Flora, and to Thomas Charles Dolphin, company secretary; Emma Elizabeth Scampton, spinster; and Gilbert Scott Ram, electrical inspector to the Home Office

1928 – Alfred Fridlander is buried in Whitton Cemetery, Birmingham

1929 – Flora Fridlander dies in Coventry on 8 Jan 1929. She leaves her estate of £1,734 12s. 6d. to their unmarried daughter Annie Esther Fridlander.

Their children:
    Annie Esther Fridlander 7 Mar 1864–1963
        Unmarried
    Adelaide Rachel Fridlander 1866–1946
        Married Gilbert Scott Ram 1865-1938. Civil Servant.
        No children
    Ernest David Emanuel Fridlander 1870–1960
        Artist & Poet.
        Married Ellen Ethel Martin 1873-1969. Artist.
        No children
    Beatrix Fridlander 1876-1964
        Artist. Married Mr. Martin. No known children

Mary Emanuel Fridlander & Michael Klean

1836 – Michael Klean born to Alexander Klean

1844 – Mary Emanuel Fridlander born to David & Esther Fridlander

1863 – Mary Emanuel Fridlander (19) and Michael Klean (27) watch manufacturer, of Middleton Street, Clerkenwell, London are married at 1, Bristol Road, Kings Norton – close to the family home of Mary's parents David & Esther Fridlander.

1871 – Michael Klean dies on 13 Nov 1871 at 36 Milner Square, Islington, London, England

Philip, Evelina & Flora Solomon

1813 – Philip Solomon born in Exeter to Jacob Solomon & Sarah Solomon née Levy

1822 – Evelina Brandon born to Abraham Pinto & Judith Brandon of the Sephardi community in Kingston, Jamaica

1842 – Philip Solomon & Evelina Brandon are married on 25 Sep 1842 in the Ashkenazi Jewish Congregation, Kingston, Jamaica

1844 – Flora Sarah Solomon born in Mandeville, Manchester, Jamaica to Philip & Evelina Solomon. Flora, later to become the wife of Alfred Fridlander, will name their home in Coventry as 'Mandeville'

1861 – Philip Solomon, watch manufacturer in Coventry. 25 Butts (now Nisa Supermarket)

1862 – Employees of Cohen, [Philip] Solomon & Co, watch manufacturers defend their employers' fair wages and their voluntary contributions to the Relief Fund for the Poor in response to accusations by the Coventry Standard

1863 – Flora Sarah Solomon marries Alfred Emanuel Fridlander – see above

1864 – 9 Jan 1864 the Globe: From Last Night's Gazette: Partnerships Dissolved:
P. Cohen and P. Solomon, Coventry, watch manufacturers

Philip Cohen & Priscilla Solomon

1820 – Priscilla Solomon born in Exeter to Jacob & Sarah Solomon

1827 – Philip Emanuel Cohen born in Prussia, son of Joseph Cohen

1846 – Philip Cohen establishes his watch manufacturing business in Coventry.

1850-53: President of Coventry Hebrew Congregation

1851 – Philip at Market Street, Coventry

1851 – Philip Cohen marries Priscilla Solomon (1824-1891), daughter of Jacob Solomon and sister of Philip Solomon, at her parents' family home: 20 Manchester Buildings, Westminster on 31 Dec

1861-81 Oxford Terrace, Hearsall Lane, Coventry

1862 & 1864 – Cohen & Solomon & Co Watch Manufacturers – See above

1871 – Employing 75 men & boys

1881 – Employing about 100 men

1891 – Priscilla Cohen dies 21 Dec 1891 at Colmar House, Warwick Road, Coventry.
Buried at London Road Cemetery. See her tomb in Jewish Cemetery section p.49ff.

1893 – Cohen watch appeared 25th in the Royal Society, Kew Observatory Annual Report of the best performing watches in the Country

1898 – Philip Emanuel Cohen of Coventry, Watch Manufacturer died 3 October 1898.
Buried at London Road Cemetery. See his tomb in Jewish Cemetery section p.49ff.

Mendel, Minah & Joseph Radges

1813 – Mendel Radges is born in Miłosławice, Gaur, Posen, Prussia

18xx – Mendel Radges marries Minah / Mena / Mennie / Minnie, presumably in Prussia

1841 – Son Joseph Radges born to Mendel & Mina in Prussia

1851 – Mendel & Minah Radges immigration to Sheffield

1855-62 – Mendel Radges and family at 4 Castle Street, Sheffield – Dealer in watch materials

1857 – Mendel Radges is robbed of a gold watch in Sheffield. Joseph Radges testifies in court.

1859 – Mendel Radges British Naturalization

1868-98 – Mendel & Mina Radges live at 11 Butts, Coventry – Now Olive Tapas & Wine Bar.

1869 – Joseph Radges marries Rachel Harris in London

1874 – Joseph Radges Treasurer of Coventry Hebrew Congregation

1875-80 – Joseph Radges President of Coventry Hebrew Congregation

1880-1891 – Joseph & Rachel Radges at Argyle House, Butts, Coventry

1881 – J. Radges applies for a patent regarding Wheels – perhaps a specialised watch wheel?

1888 – Mina Radges dies in Coventry. See memorial stone in Jewish Cemetery section p.49ff.

1896 – Joseph & Rachel Radges at Rothsay Terrace, Barras Lane, Coventry

1898 – Mendel Radges dies at 11 Butts, Coventry.
   See his memorial stone in Jewish Cemetery section p.49ff.

1901 – Joseph Radges at Stroud Green, London

1915 – Joseph Radges dies in London

Francis & Paulina Silveston

1825 – Harris Francis Silveston born in Poland

1825 – Paulina Cohen born in Prussia

1841 – Francis Silveston immigrated to Nottingham where he was a Painter & Glazier

   – Francis Silveston moves to Leicester

1849 – Francis Silveston moves to Coventry

1851 – Francis Silveston resident in Bayley Lane, Coventry

1851 – Francis Silveston marries Paulina Cohen on 18 Feb 1851 in Registry Office, Coventry

1856 – Francis & Paulina Cohen in Spon Street. Francis applies for British Naturalization.

1861 – 141 Spon Street, Coventry

1856-63 – President of Coventry Hebrew Congregation

1860's - Francis Silveston becomes a naturalized British citizen

1871, 74, 75, 80, 81, 83, 91 – Resident at 2 Spon Street, Coventry
   Occupation: Watch Manufacturer. Retired by 1891 census.

1892 – Paulina Cohen dies in Coventry. Buried in London Road Cemetery, Coventry.
   See memorial stone in Jewish Cemetery section p.49ff.

1898 – Francis Silveston dies on 30 Jan 1898 in 2 Spon Street.
   Buried in London Road Cemetery, Coventry.
   See memorial stone in Jewish Cemetery section p.49ff.

Marks Baum

1830 – Marks Baum born in Bingen en Rhine, Germany

1833 – Caroline Klean born in Bingen en Rhine, Germany. The sister of Michael Klean, watchmaker who would marry Mary Fridlander, the sister of Alfred Emanuel Fridlander

1861 – 33 Fleet Street, Coventry. Watch Material Importer, with Alfred Fridlander, Shanethe & Siemon Klean as visitors

1864-75 – 41 Spon Street, Coventry

1873-75 – President of Coventry Hebrew Congregation

1881 – Holyhead Road. Naturalised British Subject

1883-6 – Elmden Villa, Holyhead Rd.

1888 – Caroline Klean dies 8 Jul 1888 in Coventry, Buried in London Road Cemetery, Coventry. 'A loving wife, a tender mother & a true friend, her memory will be ever cherished in the hearts of all those who deeply mourn her loss.' See memorial stone on Page 49.

1891 – 12 Rothesay Terrace, Barras Lane, with Siegfried Bettmann

1907 – Marks Baum dies 11 Jun 1907 in Coventry. Buried in Whitton Cemetery, Birmingham

Levin & Jane Joel

1821 – Levin Joel born in Chodziesiesin / Cuxhaven, Prussia

1831 – Levin Joel immigration with father to N. Shields, Co. Durham

1829 – Jane Flatow/ou/au born in Leith, Edinburgh to Solomon & Maria Joseph Flatou. Both of Jane's parents were from Prussia.

1845 – Levin Joel arrives in England

1848 – Levin & Jane are married in Leeds

1859 – Levin & Jane Joel to Coventry

1861 – Maria Joseph Flatou dies in Coventry on 24 March. She is buried in Betholom Row Cemetery, 5 Ways, Birmingham as there is no Jewish cemetery yet in Coventry.

1861 – Joel family at 9 Duke Street, Chapelfields with recently widowed Solomon Flatou.

1863 – Henry Joel dies aged just 3

1867 – Jane's flamboyant brother, the art dealer Louis Victor Flatau, dies in London

1868 – Rosa Joel dies aged just 12 and is the first person buried in the Coventry Jewish Cemetery. See memorial stone in Jewish Cemetery section p.49ff.

1868 – Jane Joel contests the revised will of her brother which reduces her legacy

1870 – Levin Joel British Naturalization

1871 – Jane and the remaining children are in Hearsall Terrace, Chapelfields. Levin is out on the road as a commercial traveller back in their familiar territory of Leeds. With him are Moss Fridlander Jnr, and Lewis Anidjah – grandson of David Anidjah.

1876 – David Anidjah dies and is the second burial in the Coventry Jewish Cemetery

1877 – Levin Joel dies and is the fourth burial in the Coventry Jewish Cemetery. See p.49ff. (He is buried adjacent to Joseph Cohen, the third burial)

1881 – Jane Joel continues the watch manufacturing business, now at 4 Oxford Terrace

1885 – Jane Joel is declared bankrupt

1890 – Jane Joel emigrates to South Africa with her son Lewis Joel & his family

1896 – Jane Joel dies in East London, South Africa

Coventry Jewish Cemetery

In 1863 the Coventry Hebrew Congregation obtained approval to buy a plot of land in the London Road Cemetery – seven years ahead of the building of the Synagogue.

Several of the Coventry Jewish Watchmakers and their families are buried in the London Road Cemetery. Here are some of those memorials – somewhat overgrown as of October 2022.

Philip & Priscilla Cohen

Francis & Pauline Silveston

Now restored to view:

Philip & Priscilla Cohen. Francis & Pauline Silveston

Caroline Baum (1831-1888)    *Minna (1814-1888) & Mendel (1813-1898) Radges*

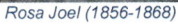

Rosa Joel (1856-1868)    *Joseph Cohen (1798-1868)*    *Levin Joel (1816-1877)*

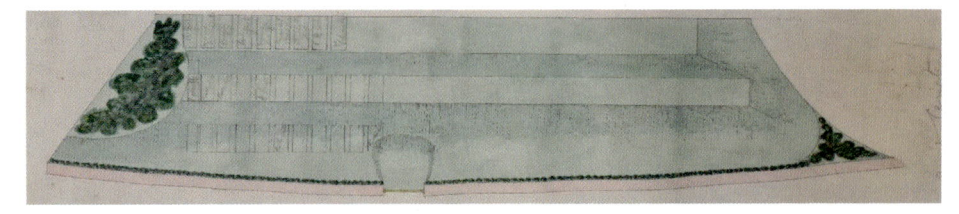

1869 Plan of the Burial Ground belonging to the Hebrew Congregation, Coventry

Other Key Coventry Watchmakers
25. Summary Overview

See the following link for a summary overview of the history of watchmaking in Coventry:
https://www.firstclasswatches.co.uk/blog/2021/08/the-history-of-watchmaking-in-coventry/

26. Bahne Bonniksen

One of the leading entrepreneurs in the expanding watchmaking trade in Coventry was Bahne Bonniksen. Although non-Jewish, Bahne was also an immigrant to England and would have had to have made many of the same adjustments to life in a new land and language as the Jewish watchmakers.

Born in 1859 in Schleswig, son of a German farmer and a Danish mother.

He was baptised in Burkal, Tonder, Denmark.

Married Fanny Durrant a Miller's daughter in Milford.

Bahne Bonniksen (revolutionized Coventry's declining watchmaking industry with his 'karussel' watch. Besides being a course instructor in horology at the Coventry Technical School he also ran his own watchmaking workshops in Norfolk Street, where he employed about 25 watchmakers.

His invention, the Karussel device, was patented in 1882. These highly precise movements were often awarded "Kew Class A" rating certificates and in general supplied to the most renowned watchmakers of the time, such as Joseph White, Smith & Sons, Robert Milne and others. Watches entirely signed by Bahne Bonniksen are exceedingly rare.

Both Karussel and Tourbillon are revolving escapements with the escapement placed on a small rotating platform, a device to eliminate errors of rate in the vertical positions. The Karussel escapement however is driven by a fourth wheel, which is not fixed but rotates within the platform. Consequently it is turning at a much slower rate than tourbillon carriages, varying from 34 to 52.5 minutes per full rotation depending upon the design. It is also more robust and easier to produce than a tourbillon, hence less expensive.

In 1907 Bahne formed the British Watch and Clock Makers' Guild with the support of the British Horological Institute.

Bahne and Fanny Bonniksen lived at 95 Tachbrook Road, Leamington with their 5 children.

Bahne died in 1935

27. Rotherham and Sons – Founded by Samuel Vale

Summary

The Vale family had been in Coventry since the late 1600's. Samuel Vale (c.1725-1786) set up his firm Vale & Sons in 1747. He was based at Spon Street, Coventry from 1776.

Vale was among the most skilful watchmakers in the city. In 1790 he hired Richard Rotherham to be his apprentice. Richard Rotherham was born Claverdon then lived Fleet Street & St John's Coventry

Rotherham quickly took after his mentor. The movements from the firm were created with high-quality materials and incredible attention to detail. After being listed as a partner in the firm, as well as years of hard work, in 1842 Vale & Sons was renamed Richard Kevitt Rotherham & Sons.

See detailed history below.

See also: https://www.vintagewatchstraps.com/rotherham.php#RHR

Detailed History

Rotherham's was originally founded by Samuel Vale. It was later passed to his one-time apprentice John Rotherham. Rotherham's became the largest of the Coventry Watch Manufacturers. Here is the detail behind the firm and its families:

Rotherham and Sons, Spon Street, Coventry. Watch and Clock Makers: https://mb.nawcc.org/threads/vale-rotherham-movement.120854/

The firm was claimed to be founded by Samuel (I) Vale in 1747. He built up a substantial business, taking on several apprentices and from 1780 partners. Several marriages between members of the partners' families helped cement the partnerships.

There were also connections with other Coventry watchmakers and Liverpool watchmakers.

Samuel Vale, son of Benjamin Vale apprenticed to John Vale, watchmaker, of Coleshill Warwickshire on 23 May 1743 may be the founding Samuel Vale.

In 1758 Samuel Vale, watchmaker of Coventry took on Edward Farmer, apprentice.

In 1760 he took on George Howlette, apprentice.

In 1765 he married Mary Carr.

1771 Samuel Vale, watchmaker was mayor of Coventry.

c 1780 Samuel Vale established a partnership with George Howlette.

c 1786 Samuel Vale died, and his will refers to the recent partnership of himself, George Howlette (of Coventry, alderman), and his son-in-law John Carr, watchmaker. By this will his widow Mary Vale (nee Carr) succeeded him in the partnership. His son Samuel (II) Vale now a minor, later succeeds his mother as partner.

30 April 1791 a loan document shows the partners as Mary Vale widow, George Howlette and John Carr, watchmakers, loan to John Whitwell.

1791 British Universal Directory Coventry, Vale Howlette Carr & Co., clock and watchmakers, Coventry.

George Howlette son of John and Sarah was baptised at Bedworth on 3 April 1745, apprentice to Samuel Vale watchmaker Coventry 1760, partner with Samuel Vale 1780, then also John Carr 1785. George Howlette watchmaker Mayor of Coventry 1784 and 1792. Universal British Directory, 1791, Coventry, Vale Howlette Carr and Co., clock and watchmakers. George Howlette died in April 1811 and was buried at Bedworth 27 April 1811.

John Carr, son of Thomas and Mary, baptised 7 March 1755, Berkswell. Married Ann Vale, daughter of Samuel (I) Vale, at St Michael Coventry 21 February 1771. By 1791 partner with (Mary) Vale and George Howlette. Liverpool St Peter, John Carr, widower watchmaker of Coventry, married Mary Platt widow.

Mary Platt nee Tarleton was the daughter of William Tarleton, watchmaker of Liverpool. Her first husband William Platt was the son of Thomas Platt, and Bridgett Kevitt, who subsequently married John Rotherham. John Carr died in Coventry 4th quarter 1838.

By 19 July 1810 Samuel (II) Vale and John Rotherham have become partners.

Indenture of David Dry, son of Thomas Dry, weaver of Coventry to Samuel Vale, George Howlette, John Carr, and John Rotherham of Coventry, partners and watch manufacturers.

Samuel (II) Vale watchmaker Mayor of Coventry 1811,1812,1813, and 1814. born January 1772, died January 1843.

John Rotherham, was baptised in Claverdon 21 May 1758, and married a widow, Bridgett Platts nee Kevitt, at St John the Baptist Coventry 6 December 1786. They had seven sons, of whom only Richard (I) Kevitt Rotherham became a watchmaker and partner. John Rotherham of Fleet St, Coventry was buried at St John the Baptist 15 February 1823.

Richard Kevitt (I) Rotherham, son of John and Bridgett, was baptised 28 March 1789 at St Michael, Coventry. 21 March 1812 he married Charlotte Carr, a minor, with the consent of her natural and lawful father, witness John Carr senior and Mary Carr.

When their first child, Mary Ann Rotherham, was born in 1813, Richard Kevitt Rotherham lived in Spon St, Coventry, and was recorded as being a Watch Manufacturer. They had two sons John Rotherham (1815 to 1875) and Richard (II) Kevitt Rotherham (1820 to 1866) who both joined the business.

Pigott in 1822 shows (Samuel II) Vale, (John) Rotherham & Son (Richard I Kevitt Rotherham) Watch and Clockmakers, Spon St, Coventry.

Samuel (II) Vale, in the 1841 census, appears to be retired, and is shown as Independent and living with his son Samuel (III) Vale, a solicitor, in Spon St, Coventry.

Coventry archives have a photograph from 1863 of four generations of Rotherhams, Richard (I) Kevitt Rotherham, his son John Rotherham, his son another John Rotherham and his son Hugh Rotherham, all of whom ran Rotherham and sons.

29 May 1867 indenture of Henry Skinner to John Rotherham and John Rotherham junior watch manufacturers. Richard (I) Kevitt Rotherham and Richard (II) Kevitt Rotherham have died, and the business now continues with the descendants of John Rotherham.

John Rotherham born June 1815 died 30 March 1875

John Rotherham born Nov 1838 died 21 May 1905

Hugh Rotherham born Feb 1861 died 24 Feb 1939, brother Kevitt Rotherham born 27 March 1864 died 26 March 1950, and brother Ewan Rotherham born Oct 1877, died 27 January 1942.

Sources & Acknowledgements

The story of the Coventry Jewish Watchmakers is a compilation of an extensive range of sources. The search was inspired by the work of Claudette Bryanston in spearheading a study and a video about immigrant communities in Coventry over the years, and Jewish Watchmakers as an example of those who gave up their homeland to pioneer a new life in a new land and language. In doing so, they contributed to building the vibrant and diverse life that is the hallmark of our city.

The most vivid source on Coventry watchmaking for anyone is to literally step back in time into the Coventry Watch Museum in the heart of Coventry's watchmaking area in Spon Street itself. Here you can become immersed in the life and lives of the watchmaking families, and the tools and products of the watchmaking trade. My thanks particularly to Leonard Heanes and Paul Shufflebotham for their invaluable assistance.

Don't just visit https://www.coventrywatchmuseum.co.uk/, go, and step into the Watch Museum itself and soak up the atmosphere in Coventry last remaining watchmaking courtyard. Currently open Tues & Sat 11:00-15:00. The museum is closed in the cold dark winter months when, somehow, the watchmakers themselves had to keep working to feed their families.

With thanks to Claudette Bryanston and the team at http://stampproductions.co.uk who's work in producing the video 'Migration Stories – Hidden Histories' inspired the research for this book.

Coventry has an excellent range of historical research books which reference the watch trade. These include:

> **The Jews of Coventry** by Harry Levine, 1970. This is the definitive reference for the history of Coventry's Jewish Community. It is written by one of its members on the occasion of the centenary of the Coventry Synagogue in which the Jewish Watchmakers played such a pivotal part.
>
> **The Coventry Watchmakers' Heritage Trail** by Andrew Barber & Jane Railton. 3rd Edition edited by J.R. Leech & Paul Shufflebotham, March 2014. This can be used as a guided tour of key watchmaking sites in Spon End & Chapelfields as well as for its wealth of research material. It is published by the Coventry Watch Museum. It includes a chapter on the life and works of Philip Cohen.
>
> **The Coventry we have Lost – Earlsdon & Chapelfields Explored** by David Fry & Albert Smith
>
> **Earlsdon Heritage Trail** by Mary Montes
>
> **A History of Coventry** by David McGrory

A rich tapestry of the developing, and then declining, watch industry can be seen through the 1851-1921 Censuses which are drawn on extensively in this work.

For key individuals and events, Birth, Marriage & Death Records have been retrieved, or their newspaper publications found.

Further Newspaper Archive articles, are referenced including the 1841 London Evening Mail newspaper article about Moss Fridlander's collaboration with Coventry's Chief of Police in capturing the Oswestry Murderer; Lizzie Baum & Jacob Landau's 1891 marriage and many more.

The Coventry Watch Manufacturers promoted themselves extensively across the country in various Trade Directories. These provide us with dates, locations and watch trade roles which contribute to the rich tapestry that is woven together here. These include:

> 1842 Pigot's Directory
> 1845 Post Office Directory of London
> 1866 Morris's Commercial Directory of Warwickshire
> 1868 Post Office Directory of Warwickshire
> 1868 Kelly's Directory
> 1872 Kelly's Directory of Warwickshire
> 1875 Francis White & Co.'s Commercial & Trades Directory of Warwickshire
> 1880 Kelly's Directory
> 1891 Where to buy in Coventry

Index

Key: **99** – **Bold**: Main reference   99 – *Italic*: Photo or graphic.

A Contrast - Coventry Standard article 1862 12
A45 37
Abraham 3
Adler, Dr. Herman, Chief Rabbi (1839-1911) . **5**
Adler, Dr. Nathan, Chief Rabbi (1803-1890). vii, 1, 14, 22, 44
Allesley Old Road 4, 36
Almack's Assembly Rooms...*See* Willis Rooms
America .... *See* USA
An Unwarrantable Attack - 1862 newspaper response 13
Anidjah Family **17, 18**
Antisemitism 1, 5, **12**
Ashkenazi 1, 46
Austria 3
Automotive Design & Manufacture...... *See* Car Design & Manufacture
Barber, Tony .... vii, viii, 25, 39
Barras Lane ..... vii, 2, 21, 29, 31, 33, 44, 47, 48
Baum Family 1, 33, 37, 48
Baum, Caroline née Klean (1833-1888) 23, 26, 27, 42, 48, 50
Baum, Lizzie 33, *See* Landau, Lizzie née Baum (1864-1939)
Baum, Marks (1830-1907) ... 19, 23, 25, 26, 27, 33, 37, 38, 42, 48
Baum, Michael (1860-) 25, 26, 27
Bavaria .... vii, 1, 14, 26, 27, 42, 43
Bayley Lane 33, 34, 47
Becket, Play by Lord Tennyson...................... 5
Benjamin, Sarah née Silveston (1852-1939) . 1, 14, 23, 25, 26, 27, 44, 46, 52
Bettmann, Siegfried (1863-1951) 33
Bicycle Making 2
Bicycles .... *See* Bicycle Making
Bingen am Rhine, Germany 26
Birmingham..... 3, 14, 19, 25, 42, 43, 44, 45, 48
Birmingham Old Road... *See* Allesley Old Road
Blessing 19
Bloody Foreigners, book by Robert Winder...vii
Bonniksen, Bahne (1859-1935)............... vii, 51
Brandon, Abraham Pinto & Judith 46
Brandon, Evelina.....See Solomon, Evelina née Brandon (1824-1896)
Breakfast 3
Britannia 5, 41, 42
British Horological Institute........................... 51
British Watch and Clock Makers' Guild ... vii, 51
Bryanston, Claudette .... viii, 55
Butcher Row, Coventry 3
Butchers Row 34
Butts, The 2, 26, 28, 29, 40, 44, 46, 47

Cambridge Villa, Holyhead Road, Coventry.14, 30, 44, 45
Car Design & Manufacture ....viii, 2, 23
Cemetery............ *See* London Road Cemetery, Coventry
Chapelfields........... vii, viii, 2, 12, 13, 17, 48, 55
Chester..37
Chief Constable .... *See* Chief of Police
Chief of Police 1, 9, 35, 55
Chief Rabbi See Adler, Dr. Nathan (1803-1890) & Adler, Dr. Herman (1839-1911)
Chirk.. 7
City .... *See* Coventry, City of
Cohen Family 1
Cohen, Isaac (1728-1835) & Judy (1722-1833)
....**3**, 5, 34, 41
Cohen, Joseph (1798-1868) 3, 43, 46, 50
Cohen, Pauline(a)... *See* Silveston Pauline née Cohen (1825-1892)
Cohen, Philip (1827-1898)....*0*, 1, vii, 1, 2, 3, 4, **10**, **12**, 18, 21, 22, 23, 25, 36, 41, 46, 49, 55
Corporation Street 29
Councillor 1, 32
Court 16 .... *See* Spon Street, Court 16
Coventry Great Fair 5
Coventry Hebrew Congregation 3, 4, **23**, 44, 46, 47, 48, 49
Coventry Herald................................ 1, 3, 4, 5
Coventry Herald and Free Press 1, 5
Coventry Herald and Observer 3, 4
Coventry Jewish Cemetery .. *See* London Road Cemetery, Coventry
Coventry Jewish Watch Manufacturers............**2**
Coventry Police ....*See* Chief of Police
Coventry Railway Station............................ vii
Coventry Ring Road ....*See* Ring Road
Coventry Standard, The... 7, 12, 13, 14, 19, 21, 22, 25, 46
Coventry Synagogue .. vii, viii, 3, 16, 17, 19, **21**, **22**, 23, 25, 26, 27, 29, 33, 44, 55
Coventry Times 1, 13, 14
Coventry Volunteer Fire Brigade....................1
Coventry Watch Museum .. viii, 2, 4, 29, 38, 39, 55
Coventry Watchmakers Heritage Trail 36
Coventry, City of ... iii, viii, 1, 2, 3, 14, 34, 37, 43
Deaths 3, 4, 9, 41, 42, 43, 45, 46, 51, 52, 53
Derby Lane 3
Earlsdon 2, 55
Electoral System..5
Emanuel, Abraham (1817-1889).................21
English language 19
Entrepreneur viii

57

Eulogy 4, 48
Evans, Emma - Victim of the Oswestry Murderer 7
Exeter 26, 27, 46
Fair Rosamund 3, **5**, 41
Families .... viii, 1, 23, 25, 31, 33, 49, 52, 55
Family Tree ....viii, 25, 26, 27, 28
Fear of 'the other' 1
Fire Brigade ....See Coventry Volunteer Fire Brigade
Flatow / Flatau Family .... *See* Joel Family
Fleet Street 3, 7, 23, 29, **35**, 43, 48, 51
Frederick the Great 3, 41
Fridlander Family 1, 29, 30, 31
Fridlander, Alfred Emanuel (1840-1928) .. *0*, vii, 1, **14**, 16, 18, 21, 22, 23, 25, 26, 27, 40, 42, 44, 45, 46, 48
Fridlander, David (1807-1876) ...13, 14, 22, 25, 26, 27, 42, 43, 44, 45, 55
Fridlander, Flora Sarah née Solomon (1845-1929) 1, 14, 26, 27, 40, 44, 46
Fridlander, Mary (1803-1858).7, 26, 27, 35, 42, 43, 48
Fridlander, Mary Emanuel...... *See* Klean, Mary Emanuel née Fridlander (1844-1907)
Fridlander, Moses (Moss) Jnr (1848-1925) . 18, 25, 26, 27, 42
Fridlander, Moss (Moses) Snr (1803-1865)1, **7**, 9, 25, 26, 27, **35**, 42, 43, 55
Germany ....vii, 1, 23, 26, 27, 43, 44, 48
Glass 19, 44
Global Leadership......... *See* World Leadership
Great Butcher Row*See* Butcher Row, Coventry
Great Fair .... *See* Coventry Great Fair
Guilty, unless proven innocent 1
Hambro Synagogue............................... 26, 27
Harris Family.. **28**
Heanes, Leonard 4, 38, 55
Hearsall Lane..................................... 4, 36, 46
Hertford Place......................... 1, 14, 29, 32, 45
Holyhead Road ... 14, 23, 29, 30, 31, 44, 45, 48
Hundt, Gillian, Professor Emeritus v
Immigrant Families.......................... iii, viii, **1**, 2
Immigrants .... iii, viii, 1, 2, 55
Immigrants, Being misunderstood................. 1
Immigrants, Benefits to Society of iii, viii, 1, 2, 5
Immigrants, Homeland memories 1, 32
Immigration ....*See* Immigrants
Industrial Heritage....... *See* Heritage, Industrial
Inevelyan Hotel, Leeds 18
Irving, Henry 5
Jamaica 1, 14, 26, 27, 32, 44, 46
Jewish Chronicle...................................... 3, 18
Jewish Community...................... viii, 1, 23, 55
Jews of Coventry, Book by Harry Levine (1970) ....viii, 3, 4, 16, 23, 55
Joel Family..............................**17**, **18**, **28**, 36, 50

Justice of the Peace 1, 14, 32
Karussel watch movement.............................51
Kenning Illustration & Creative Design..........***39***
Kew Observatory . vii, viii, 11, 15, 44, 45, 46, 51
Klean Family... 1
Klean, Caroline *See* Baum, Caroline née Klean (1833-1888)
Klean, Mary Emanuel née Fridlander (1844-1907) 26, 27, 45
Klean, Michael (1836-1871). 25, 26, 27, 42, 43, 44, 45, 48
Land Registry 21
Landau, Jacob (1860-1957)........ 19, 26, 27, 55
Landau, Lizzie née Baum (1864-1939) .. 19, 26, 27, 55
Leeds 18
Lichfield 37
Loew's Dancing Show 5
London .viii, 1, 2, 3, 7, 9, 23, 26, 27, 28, 37, 42, 44, 45, 46, 47, 48, 49, 55
London Road Cemetery, Coventry...viii, 17, 18, 46, 47, 48, **49**
Lord Tennyson 5
M6 37
Mandeville, Home of Alfred & Flora Solomon 1, 14, **32**, 45, 46
Mandeville, Jamaica................ 1, 14, 26, 32, 46
Marriages .. viii, 1, 3, 14, 19, 23, 25, 41, 42, 45, 46, 52, 53, 55
Mayor of Coventry 33, 52, 53
Medieval wooden framed buildings...............37
Minyan...19
Misunderstanding Immigrants........................1
Moss Fridlander *See* Fridlander, Moss (Moses) Snr (1803-1865)
National Physical Laboratory 15, 45
Naturalisation 1, 43, 48
New Synagogue, London 26, 27
Old Coventry Jewry 3, 5
Oswestry Murderer 1, 7, 35, 43, 55
Oxford Terrace, Hearsall Lane.....vii, 29, 36, 46
Philip Emanuel Cohen Charity 21
Pioneers .... viii, 2, 23, 55
Pocket Watch, Philip Cohen 4
Poland 1, 47
Police ....See Chief of Police
Poole, Benjamin 1852 History of Coventry......3
Prescot, near Liverpool..................................2
President of the Coventry .. 3, 4, 23, 44, 46, 47, 48
President of the Coventry Hebrew Congregation 23
Priory Row, Coventry................................ 3, 41
Prize 14
Product Quality.. viii
Prosser, Thomas Henry - Chief Constable of Coventry 7, 8, 9

Prussia......... vii, 1, 3, 17, 26, 27, 28, 41, 46, 47
Prussian Army 3, 41
Queen Victoria .... vii, 14, 22, 35, 43
Queen Victoria Road.................................... 29
Queens Road... 29, 32
Rabbi, ChiefSee Adler, Dr. Nathan (1803-1890) & Adler, Dr. Herman (1839-1911)
Racism........................1, *See* also Antisemitism
Radges Family......................... 1, 25, 29, 40, 47
Radges, Augustus (1855-1888) 26, 28
Radges, Ernest (1876-1927).................. 26, 28
Radges, Joseph (1841-1915) 22, 23, 26, 28, 47
Radges, Louis Henry (1873-1948) 26, 28
Radges, Mendel (1813-1898).....25, 26, 28, 40, 47, 50
Radges, Minah (1814-1888) 26, 28, 50
Radges, Rachel née Harris (1845-1927) 26, 28
Radges, Rachel née Harris (1875-1951) 26, 28
Rating of Watches............................15, 44, 45
Reform Act 1832....................................... 5, 41
Representation 5
Ribbon Making viii
Ring Road....................... 14, 29, 31, 32, 37, 40
Rotherham, John (I) (1758-1823)................. 53
Rotherham, John (II) (1815-1875)................ 53
Rotherham, Richard Kevitt (I) (1789-c.1865) 53
Rotherham, Richard Kevitt (II) (1820-1866).. 53
Rotherhams Watch Manufacturers...29, 30, 31, 37, 51, 52, 53
Rothesay Terrace, Barras Lane .26, 28, 29, 33, 48
Royal Family vii
Royal Society 15, 44, 45
Russia... vii
Sabbath Class vii
Samuel, Selim (1853-1929) .... vii, 25
San Francisco.. 26, 28
Semi-detached Houses............................ 2, 31
Sephardi 1, 46
Sewing Machines.................................... viii, 2
Sheffield..26, 27, 28, 47
Shopkeepers... 5
Shufflebotham, Paul.................................... 55
Shul .... *See* Coventry Synagogue
Shulte, Moritz... 33
Silesian Wars 3
Silver Caddy Spoon 8
Silveston Family..................................... 1, 29
Silveston, Francis (1825-1898) ..21, 23, 25, 26, 27, 33, 37, 38, 47, 49
Silveston, Pauline née Cohen (1825-1892). 25, 26, 27, 49
Silveston, Sarah....... *See* Benjamin, Sarah née Silveston (1852-1939)
Slawson, Joseph - Accomplice of the Oswestry Murderer.. 8, 9

Solomon Family............................ 1, 27, 29, 40
Solomon, Evelina née Brandon (1824-1896) .1, 26, 27, 40, 46
Solomon, Flora Sarah..... See Fridlander, Flora Sarah née Solomon (1845-1929)
Solomon, Jacob (1783-1859)............26, 27, 46
Solomon, Philip (1813-1885) . 1, 12, 13, 25, 26, 27, 40, 44, 46
Solomon, Priscilla (1820-1891).........26, 27, 46
Solomon, Sarah née Levy (1790-1875) .26, 27, 46
South Africa.. 18, 28
Spon End...3, 55
Spon Street viii, 2, 3, 23, 29, 30, 31, **37**, 38, 39, 47, 48, 51, 52, 55
Spon Street, Court 16.....................................3
St. Mary's Hall 22
Stamp Productions viii
Swimming Competition 14
Switzerland...2
Synagogue .... *See* Coventry Synagogue
Tasmania..9
Terraced houses..2
Terry, Ellen..5
Theatre..5
Thunderstorm 19
Tourbillon watch movement.........................51
Trail.. viii
Transportation 9
Triumph Cycle Company 16, 33
Trunk Maker 3
USA..2, 26, 28
Vale, Samuel (I) (1725-1786).................51, 52
Vale, Samuel (II) (1772-1843).................52, 53
Volunteer Fire Brigade See Coventry Volunteer Fire Brigade
Vote...5
Watch Museum, Coventry See Coventry Watch Museum
Watch Rating....... *See* Rating of Watches, *See* Rating of Watches
Watch, Alfred Fridlander........vii, viii, **15**, 16, 25
Watch, Philip Cohen **4**
Watchmaking...**39**
Watchmaking, Decline 1, 2, 23
Watchmaking, Growth 1, 2
Weaving viii
Wedding ....See Marriages
West Cost Mainline......................................37
Williams, John - The Oswestry Murderer 7, 8, 9
Willis Rooms..1, 14
Winder, Robert vii
Wine 19
Wool Trade.. viii
World Leadership .... viii, 12
Wrexham..8